GunTrucks

A VISUAL HISTORY OF THE U.S. ARMY'S VIETNAM-ERA WHEELED ESCORT PLATFORMS

by David Doyle

Published by
Ampersand Group, Inc.
A HobbyLink Japan company
235 NE 6th Ave., Suite B
Delray Beach, FL 33483-5543
561-266-9686 • 561-266-9786 Fax
www.ampersandpubco.com • www.hlj.com

Acknowledgements:
The author gratefully acknowledges the help and resources provided by Tom Kailbourn; Jeff Houghton; Richard Killblane, US Army Transportation Corps historian; David Hanselman and the staff of the US Army Transportation Museum; Ken Whowell, Scott Taylor, the staff of the Rock Island Arsenal Museum, Dana Bell and Paul Kopsick, Historian of the National Dusters, Quads and Searchlights Association, provided previously unpublished photos and information. Special thanks go to my wife Denise, who as always has been a constant source of encouragement.

Unless otherwise noted, the author took all photos of Eve of Destruction on display and all wartime photos were provided by the US Army Transportation Museum, Fort Eustis, Virginia.

Above: The two extremes of gun truck configuration; a ¹/₄-ton M151 and, in the background, a 5-ton M54 6x6.

Cover: The gun trucks used in Vietnam where created by the men who used them and each was as unique as the GIs who served on them.

Title page: "Snoopy" was one of the breed of gun trucks developed to protect supply convoys during the Vietnam War.

Right: Arguably the heaviest armed gun trucks were operated by the four "Quad-50" batteries of air defense artillery.

Table of Contents

The 2½-ton 6x6 3

The 5-ton 6x6 8

The M151 14

The M37 17

The M55 "Quad" 19

The Birth of the Gun Truck 30

Weapons 34

The 2½-ton Gun Truck 40

The 5-ton Gun Truck 54

APC-bodied Trucks 80

Engineer Gun Trucks 88

Armored M151s 90

Armored M37s 95

The Last Survivor 97

Military Transport

Few people outside the military recognize the enormity of the logistical effort required to keep an army in the field provisioned with food, ammunition, clothing, medical supplies and equipment and a host of other wares.

While trains, ships, aircraft and helicopters form links in this supply chain, for much of the 20th century, the final link in this chain has fallen to the truck. Lengthy truck convoys snaked across Europe, China-India and North Africa during WWII. While helicopters saw some use in Korea, and their usage increased manyfold in Vietnam, it was still the truck that moved that bulk of the materials from ports, supply depots and airports to forward bases. At the time the U.S. servicemen began to arrive in Vietnam, the mainstays of the American tactical vehicle fleets were the M35 2 1/2-ton and M54 5-ton 6x6 trucks.

The 2½-ton 6x6

The M35 was a dual-wheeled derivative of the M34 which itself rode on six 11.00-20 tires. Lansing, Michigan-based Reo Motors, Inc. designed the M34, with the pilot models being tested in 1949. While the single 11.00-20 tires provided superior off-road performance, dual-wheels were preferred for on-road operation, and the M35 was equipped with dual rear wheels accordingly. The slightly smaller 9.00-20 tires permitted the elimination of the wheel wells from the bed, allowing the cargo bed floor to be flat, easing loading of the truck.

During the Korean War, U.S. strategic planners were concerned that the conflict would escalate into World War Three, and began an arms buildup and strategic industrial planning accordingly. It was felt that the Lansing firm lacked the industrial capacity to meet military's needs for the new 2½-ton truck. South Bend, Indiana-based Studebaker was struggling financially at the time, and adding that firm as a second production source both bolstered that area's economy as well as providing the military with the additional production it desired. Studebaker was awarded a contract 1950, only a few months after Reo's June 1949 contract award.

Equipped with ten 9.00-20 tires with duals on the rear, the M35 also dispensed with the wheel wells in the cargo bed, creating a flat loading floor. Like the M34, Reo and Studebaker and Studebaker's successor companies built the M35. (TACOM LCMC History Office)

Whether produced by Reo or Studebaker, or M34 or M35 variant, the Reo-designed 6-cylinder gasoline engine dubbed the Gold Comet powered these trucks. This engine was a successful commercial product for Reo, being used in many over the road trucks. The model number assigned was OA-331, and the engine was actually demonstrated to the military on 1 April 1949. Unable to keep pace with engine production for two plants, plus spares, Reo granted a license to Continental Motors to produce the OA-331 for military use only. The Continental version of the engine was the COA-331. Both the M34 and M35 2½-ton cargo trucks were part of a family of vehicles known by the army's Standard Nomenclature List as G-742. Approximately 101,000 G-742 series trucks with gasoline engines were built, encompassing a variety of body types: cargo, tanker, van and a vast array of specialized bodies.

The U.S. Military had long desired an engine that was capable of burning a variety of fuels. The ability to burn gasoline, Diesel, jet fuel, or a combination of these was advantageous from both the supply chain standpoint as well as the capability of utilizing whatever fuels were locally abundant in a given theater of operations.

In the 1950's Continental Motors Corporation's engineer Carl Bachle learned of a new engine that had been developed by M-A-N. Dubbed the "whisper engine," the new engine had been designed by Dr. S. J. Meurer to combat the prevailing (at the time) reputation of M-A-N engines as the loudest on the market. Bachle made two trips to Germany to learn how the new engine design reduced noise, and in the course of his examination, concluded that the "whisper engine" was also capable of multi-fuel operation.

Bachle struck a licensing arrangement between Continental and M-A-N for the latter firm's "whisper engine" design which utilized a combustion process that Continental would later dub the Hyper-Cycle combustion process. The M-A-N engine performed to specification, and Continental Motors redesigned their TD-427 6-cylinder truck engine into a Multifuel engine, the LDS-427. In 1958 the military contracted for 22 M35E7 test trucks powered by the new engine.

This engine was a straight six model LDS-427-2 Multifuel engine, with 427 cubic inch displacement. These engines are able to burn Diesel, jet fuel, kerosene, or gasoline, or any combination of these,

Designed by Reo Motors of Lansing, Michigan, the 1949 M34 2 1/2-ton 6x6 truck rode on six 11.00-20 tires. The M34, produced by both Reo and Studebaker into the mid-1950s, had largely been withdrawn from front-line service prior to large numbers of U.S. troops being committed to Vietnam. (TACOM LCMC History Office)

In addition to the three Transportation Groups mentioned above, Air Defense Artillery units, MP and Engineer units as well as the Marines created their own gun trucks.

The 1953 M35E5 introduced the concept of the drop-side cargo bed, which permitted rapid handling of cargo and equipment by forklifts. Regrettably, it would be over a decade later before the dropside cargo beds were produced in large quantities, meaning most of the trucks in Vietnam lacked this feature. (TACOM LCMC History Office)

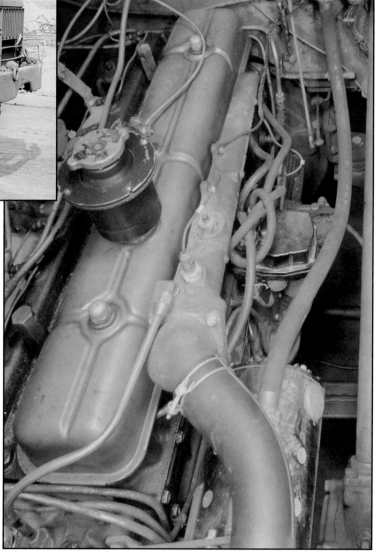

without adjustment or modification. After extensive tests this engine was adopted and installed in the G-742 series trucks. The cargo truck version of the G-742 series powered by this engine was designated M35A1 and was introduced in June of 1963. Like the gasoline-engine powered version of the cargo truck, the exhaust pipe on this truck was horizontal, exiting over the right side tandem axles.

After 16,070 trucks were built with the LDS-427-2 engine that powerplant was replaced with the LD-465 series of Multifuel engines in June 1965. The cargo trucks powered by this engine were designated the M35A2. The LD-465 had a slightly larger displacement (478 cubic inches) and was more robust. These engines were naturally aspirated, with the exhaust exiting above the right side tandems as they had on previous trucks. With the introduction of the LD-465-1C engine in later production vehicles a vertical exhaust began to be used. The exhaust on these trucks flowed from the engine to the muffler beneath the cab, then forward and up a $2^{1}/_{2}$" diameter vertical exhaust pipe extending through the rear of the right front fender. About 125,000 trucks were built with this engine.

Just before U.S. troops were withdrawn from Vietnam the LD-465-1C was replaced with the turbo-supercharged LDT-465-1C. The turbosupercharger was not added to this engine to increase horsepower, but rather to comply with EPA restrictions. As the earlier engines wore out, they were to be replaced with the LDT. The LDT-465-1C in turn was replaced with the LDT-465-1D. Trucks

with the LDT-465 engines also featured a vertical exhaust, but with no muffler in the system; the exhaust merely turned up beneath the right front fender and exited through a hole in the fender. This exhaust pipe is $3^{1}/_{2}$ inches in diameter.

Units in Vietnam

While the $2^{1}/_{2}$-ton and 5-ton cargo trucks described previously could be found in the motor pools of virtually every army unit in Vietnam, the bulk of the logistics work was done by specialized transportation units. These included the 8th Transportation Group (Motor Transport), 48th Transportation Group (Motor Transport) and the 500th Transportation Group (Motor Transport). Subordinate to these units were the 6th and 7th Transportation Battalions (Motor Transport) under the 48th Group; the 27th, 54th and 124th Transportation Battalions (Motor Transport) subordinate to the 8th Group; and the 36th and 57th Transportation Battalion subordinate to the 500th Transportation Group.

The 48th Transportation Group arrived in Vietnam from Fort Eustis, Virginia on 8 May 1966, and although technically remaining in country until 13 June 1972, the unit was reduced to zero strength in

A Reo-designed OA-331 6-cylinder gasoline engine powered both the M34 and M35. A successful engine in Reo's commercial line, the military version featured a sealed, waterproof ignition system and carburetor as well as a military standard oil filter. (Mack Museum)

Top left: The military had wanted multifuel capabilities in its tactical vehicle fleet since WWII. In 1958 testing began on the M35E7, powered by the Continental LDS-427 Multi-fuel engine, the first truly promising proposed engine with the desired capabilities the U.S. military had seen. **Top right:** The M35E7 sported a backward-sloping hood in order to accommodate the test powerplant. Ironically, the backward-sloping hood would return in the mid-1990s with the introduction of the purely Diesel-powered M35A3. **Above left:** The first production model of the Multifuel-powered cargo truck was the M35A1. Refinements in the powerplant installation permitted a conventional hood slope to be used. The right-side engine compartment panel was altered to accommodate the mushroom-shaped air intake. The exhaust was above the right-side tandem drive, in the same location it had been on the gasoline-powered trucks.

Above right: The 6-cylinder LDS-427 was equipped with a turbosupercharger, and had a 427 cubic-inch displacement. It could burn gasoline, kerosene, Diesel fuel, jet fuel and certain other fuels, or any combination thereof, without alteration or adjustment. Pure gasoline lacked the lubricating properties of the other fuels, and continued use of straight gasoline would shorten the life of the fuel injection pump.

Mid-production M35A2 trucks introduced a vertical exhaust, as on this winch-equipped model. Exhaust gasses were routed to a large muffler mounted beneath the passenger's side of the cab, then forward again and up the exhaust stack which protrudes through a hole in the right front fender. Only a few of these trucks with vertical exhausts saw service in Vietnam.

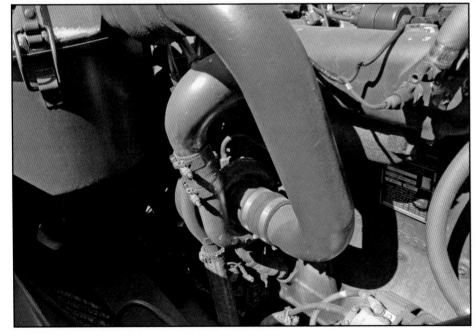

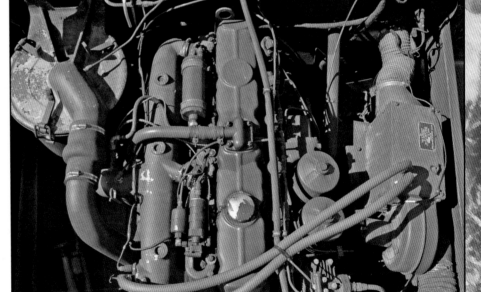

Top left: Problems with the engine in the M35A1 led to the development of the M35A2, an example of which is seen pulling a trailer off of LST-664 in Saigon in August 1968. Externally virtually indistinguishable from its predecessor, the early M35A2 was powered by a LD-465 engine, and maintained the rear, over-tandem exhaust. (NARA) **Top right:** The final production version of the M35A2 was powered by the turbosupercharged LDT-465 Multifuel engine. The turbosupercharger unit on this engine produced only a negligible increase in horsepower, but rather was added to clean up the exhaust, a requirement of new EPA regulations. A vertical exhaust was retained, albeit of larger diameter, but no muffler was used.

Above left: The addition of the turbosupercharger naturally caused a rerouting of the intake airflow as well as the exhaust. The turbosupercharger was the Schwitzer model 3LD305 or 3LJ319. **Above right:** The box at the extreme right in this view of a M35A2 engine bay contains the heater core. Heaters could be either factory or field installed, but were found on few trucks, especially in Vietnam.

The 5-ton 6x6

The International Harvester M41 was the first of the M-series 5-ton 6x6 trucks. The M41 rolled on six 14.00-20 tires. The cargo bed had wheel wells in the floor to accommodate the large tires. The trucks of this type were given the Standard Nomenclature List (SNL) number of G-744. (Patton Museum)

The five-ton 6x6 of the Vietnam era, typified by the M54A1 cargo truck, followed a development course parallel to that of its smaller sibling, the 2½-ton 6x6. The genesis for the postwar 5-ton 6x6 can be found in the June 1945 Cook Board Report. The Cook recommendations were affirmed by the November 1945 Stilwell Board Report, more properly known as the War Department Equipment Board, chaired by General Joseph "Vinegar Joe" Stilwell. This board recommended the 5-ton size with power steering as a replacement for the wartime 4-ton 6x6s typified by the Diamond T 969. Both of these boards intended the design to have a 5-year life span and the 1950 Army Equipment Board anticipated that cross-country carriers based on the T-51 design would replace the design.

The characteristics for the "interim" design, which would be given the military Standard Nomenclature List designation "G-744," were specified in MCV 204, "Military Characteristics of Vehicles, Truck, 5-ton, 6x6." The actual solicitation for bids for these vehicles, which were built to conform to Joint Army-Navy Specification JAN T-712, has not yet been located. However, located in the Reo archives are quotations to a number of manufacturers concerning the "Reo" cab. The earliest of this correspondence is with the Federal Motor Truck Company, and is dated March 24, 1950. Among the truck builders receiving quotations from Reo were Four-Wheel Drive

Auto Co. (FWD), White Motor Co., Marmon-Herrington, Federal Motor Truck Company, Ward LaFrance, and Corbitt, and International Harvester Company.

On May 2, 1950 International Harvester requested a cab quotation, which was given on May 8 at $408.86 net each FOB Lansing. This was for a prime finish cab, less boxing and crating. This figure included $143.55 for purchased items such as windshields, windshield wipers, mirrors, seats and soft top ($25.65) and windows and regulators.

On June 5 and 7, 1950 a board of officers, including Col. Joseph Colby, Lt. Col. George White, and Lt. Col. John Cone, met to discuss the procurement of a new series of 5-ton trucks. Absent was the board president, Col. David Crawford. Advisors to the board were R. F. Brown and M.C. Morrison with the Detroit Arsenal, and C.R. Tobey with the Detroit Ordnance District. The urgent requirement for these vehicles required that the contract be issued and the vehicles placed into production before tests were conducted on the

pilot models. Trucks that would later be standardized as the M41 cargo truck and M51 dump truck were both under consideration, although the cargo truck received the greatest emphasis.

Due to price factors, the vehicles under consideration were immediately confined to proposals from White Motor Company, Mack Manufacturing Company and International Harvester.

Among the factors that weighed into these discussions were engineering features, industrial mobilization factors and procurement cost.

With those factors in mind, the Mack vehicle was eliminated as their vehicle had the least desirable engineering characteristics, highest cost of those

remaining, and no offsetting industrial mobilization advantage.

International's lead-time for production expansion was a mere 30 days in the event of war, compared to the 180 days forecast by White. Further, White's sole plant was in Cleveland, whereas International had their Fort Wayne (Indiana) works available, as well as Springfield, Illinois and Memphis, Tennessee. From an industrial mobilization standpoint this was a tremendous advantage.

The International vehicle was to be powered by the 602 cubic inch R6602, which Continental Engines had previously tooled for a production of 3,000 units per month. White's engine was tooled at only 1,000 per month.

While the White design offered no advantages over the International, the IH truck was 500 pounds lighter, had 17% greater torque, 10% greater horsepower, and exceeded every area of the specification by a comfortable margin.

White quoted a price of $10,183,737 for 910 vehicles, including preproduction models, while the International Harvester price on the same package was $10,367,567.00.

Despite the $184,000 price advantage of the White proposal, the merits of the IH design and organization swayed the board, which on June 7, 1950 unanimously recommended contract DA-20-018-ORD-9197 be awarded to International Harvester Co. This contract was for 932 trucks, with a follow-on order of 10,617 trucks being placed on 4 December of that year. Also on 4 December a contract for 3,581 duplicate trucks was award to Chicago-based Diamond T Motor Car Company, and this was followed with a 29 June 1951 order with Mack Trucks for 1,665 units. Production began at International's Ft. Wayne works in January 1951. International, Mack and Diamond T would continue to receive orders for these trucks throughout the decade, until ultimately 48,639 trucks had been built with the Continental

Introduced immediately after the M41, the M54 featured dual rear wheels, and all ten of its tires were size 11.00-20. The smaller, but dual, wheels negated the need for wheel wells in the bed of the truck, permitting the vehicles to have a more cargo-friendly flat bed floor. (Patton Museum)

R6602 gasoline engine. International Harvester would build 28,089 of the trucks, with Diamond T assembling 13,591 and Mack 6,959.

Like the 2½-ton M35 Reo, front axle engagement was automatic, using an overrunning clutch. Unlike the 2½-ton truck, which used a mechanical linkage to shift the sprag unit from forward to reverse, the 5-ton used an air cylinder to accomplish this shift (which is not the same system as was used to engage the front axle pneumatically as on later versions of the M35).

Personnel at Aberdeen Proving Ground conducted testing of the new trucks in March 1951, but it was March 1953 before these vehicles were classified Standard A. Despite the hasty procurement process, the trucks were found to be basically satisfactory, a fact borne out by the basic design's continued production by various manufacturers until the 1980s, in spite of an originally planned 5-year life span.

One of the first major modifications to the five-ton's basic design was to look for an alternate powerplant. The gasoline-burning Continental R6602 was powerful, but extremely thirsty, with the mileage on heavily laden examples hovering around 2 MPG.

Testing of a commercial vehicle powered by a Mack ENDT-673 engine by the Army in the late 1950s revealed that there was considerable potential for a Diesel-powered tactical truck. As a result of these tests, in early 1959 OTAC (Ordnance Tank Automotive Command) recommended that a number of the ENDT-673 engines be obtained for evaluation in the M54 6x6 truck. Accordingly, seven M54 cargo trucks were repowered with the Mack ENDT-673 engine in February 1960. With the Mack engine installed, these test trucks were designated M54E3.

The ENDT-673 was a turbosupercharged 6-cylinder valve-in-head water-cooled compression-ignition (Diesel) engine generating 211 gross brake horsepower at 2,100 revolutions per minute. At the same time the Diesel was installed, the Spicer 6352 five-speed manual transmission was replaced in favor of a 6453, also a five-speed, but with gear ratios more befitting of the slow-turning Mack engine.

Concurrently, seven other trucks were repowered with a Cummins Model C200-A turbosupercharged compression-ignition engine, developing 200 gross brake horsepower at 2,800 revolutions per minute. These seven trucks were designated M54E4. Both the M54E3 and M54E4 underwent testing at Aberdeen Proving Ground, Maryland, Yuma Test Station, Arizona, and Fort Greely, Alaska. While both the M54E3 and the M54E4 met all standard 5-ton truck performance specifications the general impression was that the M54E3 was slightly superior and in June 1962 it was classified Standard A and designated M54A1.

In a program that began immediately and continued until 1963, M52 tractor and M54 cargo trucks were converted to the Mack ENDT-673 Diesel engine. This was a joint project between Mack Trucks and the Diamond T Motor Truck Company. Diamond T replaced the gasoline engine in 5,356 existing cargo and tractor trucks with the Mack engine. The vehicle model identifications for these trucks when equipped with the Diesel engine were M52A1 and M54A1 respectively.

The installation of the ENDT-673 was short-lived. After only a year it was decided to use Multifuel engines wherever possible in the

The Continental R6602 six-cylinder gasoline engine powered the early trucks of the G-744 type. This engine, part of Continental's commercial line, had a 602 cubic inch displacement and developed copious amounts of torque, but at the price of pitiful fuel economy. (Patton Museum)

tactical vehicle fleet. For the five-ton, the engine chosen was the LDS-465-1A. The LDS-465-1A was a turbosupercharged version of the LD-465 used in the 2½-ton trucks. This engine featured additional rings, improved oiling and a different type of fuel injection nozzle. The turbosupercharger on the LDS-465 provided a boost that, along with the different injector nozzles and altered fuel injection pump settings, increased horsepower. This was opposed to the small turbo on later 2 1/2-ton trucks with LDT engines, where the turbo was used merely to clean up the exhaust. However, most of the other parts were completely interchangeable between the LD-465 and LDS-465, easing the supply chain. With the Multifuel engines installed, the model suffixes of the cargo and tractor trucks changed to A2.

In 1963 Studebaker was awarded the contract to build the Multifuel version of the G-744 5-ton trucks. Initially the contract

was for 4,159 units with contract extensions bringing the total to 9,343 vehicles. However, before production of the trucks could begin, in February 1964 Kaiser-Jeep bought Studebaker's Chippewa Avenue truck plant, and the G-744 contract. The Army approved the contract transfer the following month. From June 1964 through March 1968, contracts totaling 44,490 Multifuel-powered 5-ton 6x6 trucks in all body types were awarded, and all would be produced in the South Bend plant. On March 26, 1970 Kaiser-Jeep became the Jeep Corporation, and the South Bend facilities were part of the General Products Division. Just over a year later, on March 31, 1971 the General Products Division was spun off to form AM General, although by that time the M54A2 was no longer in production.

Troops in Vietnam preferred the Mack engine over the Multifuel, feeling it was more powerful and more reliable.

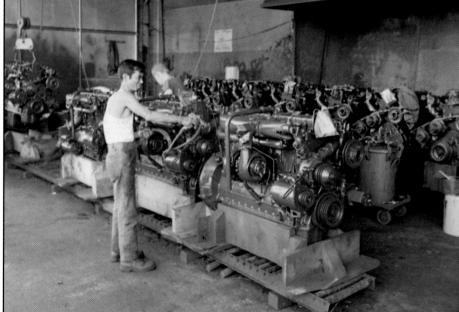

Top left: In 1959 the Army began looking in earnest at Diesel power-plants for the five-ton 6x6 truck family. The M54E3 shown here was one of 22 test vehicles trialed with the Mack ENDT-673 engine. **Top right**: The ENDT-673 was widely used in commercial applications. The 22 Mack-powered trucks were tested against a like number of trucks powered by the Cummins C-200A engine, with the Mack winning the competition by a narrow margin. (Patton Museum, both) **Above left**: Use of the ENDT-673 was short-lived, as just over a year after cargo trucks with that powerplant were standardized as M54A1, the M54A2 was introduced. The M54A2, shown here, is powered by a LDS-465 Multifuel en-gine, and features a vertical exhaust rising through the right front fender. **Above right**: The LDS-465 provided the army with the desired multifuel capability in the five-ton chassis, but the LDS lacked the reliability of the Mack ENDT. This led the military to establish in-the-ater engine rebuild depots such as this one on Okinawa. (NARA)

The LDS-465 shared many parts with the LD-465 and later LDT-465, but utilized a different turbosupercharger and fuel injectors in order to generate the additional horsepower required by the larger trucks and its increased load capacity. (US Army Ordnance Museum)

Engineer units were furnished with the M51 dump truck, a five-ton 6x6 of the G-744 series, powered by the R6602. Later models were the M51A1 and M51A2 trucks, with the Mack and Multifuel powerplants respectively. (International Harvester)

Engineer bridging units in Vietnam utilized this M328 bridge truck, built on the M139 chassis, to transport M4T6 bridge components. With an overwidth bed and dual 14.00-24 tires requiring axle spacers, the resultant truck was a formidable-appearing vehicle. The R6602 gasoline engine powered all of them. Some units adapted these trucks, once the bridge was unloaded, for use as gun trucks. (International Harvester)

The M151

The M151 was functionally the Vietnam-era equivalent of the WWII Jeep. Dubbed the MUTT, these ¼ ton 4x4 vehicles were built in three base models, M151 (shown here) M151A1 and M151A2, as well as several specialized versions of each. The M151 was developed by Ford, but was also built by Willys and Kaiser-Jeep, who built this example. (NARA)

The use of ¼-ton 4x4 utility vehicles can be traced back to the WWII-era jeep. That vehicle, the Willys MB and Ford GPW, evolved into the M38 by the late 1940s. The M38, though featuring such standardized M-series components and 24-volt electrical system and instrumentation, was nevertheless little advanced from its predecessors.

Looking for a vehicle with notable improvements in capabilities, the Army, through Ordnance Committee Minutes 33521 on 4 January 1951, began work toward a new ¼-ton truck, to be designated T122. To carry out this work, Ford was awarded contract DA-20-18-ORD-11283, with a value of $175,000. A short time later, Ordnance Committee Minutes 33850 redesignated the T122 the XM151. Following several years of testing and numerous modifications in the design, in 1958 the M151 was standardized and on 25 June 1959, almost nine years after Ford initiated work on the project, the first production contract, for 4,050 vehicles, was awarded. Despite competitive bids by others, the bid by Ford represented the best value on this contract, which was then awarded to the company, and deliveries were scheduled to begin in March 1960.

Not unlike its predecessors the M38 and M38A1, the M151 had the military standard 24-volt electrical system, and selective two- and four-wheel drive. However, unlike the older designs, the M151 used a unibody design, had a single-speed transfer case, and significantly featured all-around independent suspension. A four-speed transmission was used, whereas the MB/GPW/M38/M38A1 all had three speed transmissions. By use of low-reduction first and reverse gears, Ford's engineers were able to eliminate the need for a separate low range in the transfer case.

Two years following the initial contract another contract, DA-20-018-ORD-23240, was awarded, this for 13,124 of the vehicles. These vehicles were to be produced at Ford's Livonia, Michigan facility.

Yet another contract, DA-20-019-ORD-3941, was awarded in May of 1962. This time, a different bidder prevailed, with Willys Motors being awarded the contract for the production of 14,625 M151 trucks. In December 1962 Willys was also awarded a second contract, DA-20-113-AMC-0846 (T), for an additional 9,883 ¼-ton vehicles. This, the second contract, was modified after award to reflect the introduction of a new model designation, the M151A1. Although not receiving these production contracts, Ford remained the lead design firm.

The change of designation to M151A1 reflected a redesigned rear suspension. The initial rear suspension design sometimes buckled or collapsed, particularly when burdened with mounted weapons and cargo. The A1 introduced high-strength rear suspension arms with extra bump stops. Production of the M151A1 began in December 1963, the truck data plates bearing the manufacturer's name of Kaiser-Jeep, reflecting a corporate change occurring in March 1963.

Ford regained production status when on 16 February 1964 the firm won the first multi-year production contract for the MUTT (DA-20-113-AMC 02787 [T]). In particular, the multi-year award made this contract appealing to Ford. Despite this, Ford tendered their bid over the protestations of Ford's automobile and truck divisions, who feared adding this vehicle, with no civilian counterpart, to the production lines could tie up capacity needed to fill the civil market. Thus, production of the Ford M151A1 became the responsibility of the Industrial Division. Production began at the Highland

Park Tractor Plant during the last two months of 1964 and continued into 1969.

Despite the improvements offered by the M151A1, the government was becoming increasingly concerned about the operation of the M151 and M151A1, which were involved in an inordinately high number of accidents. At least initially, the military listed the cause as "driver error," but as the accident reports continued to pile up, it became evident that there was something else at fault.

Investigation showed the root cause of the problem lay with the fact that handling qualities of the vehicles with their independent suspension were very different from anything else in the army's tactical vehicle fleet, or from what most soldiers had become accustomed to in civilian life.

Due to the independent suspension and the lack of body roll, the M151 and M151A1 felt stable when cornering, right up to the point of rollover—the driver did not sense any danger until it was too late. Further, as was standard practice by the military, the bulk of the initial testing of the M151 design had been done with fully laden test vehicles. A fully-loaded M151 has considerably lower tipping tendency than an empty vehicle, thus the severity of the problem was masked until the vehicle was in use by troops in the field, who tended to operate the vehicle with less than a full load.

Records indicate that in fiscal 1967 there were 3,538 accidents involving the M151. From these, there were 104 fatalities and 1,858 injuries. In an effort to stem this loss of life, the U.S. Army introduced training courses, training films, circulars and even went so far as to mandate that no one drive the M151-series unless their SF-46 (Driving Permit) had been specially endorsed.

Despite these efforts, the accident rate remained high. An effort was then made to redesign the vehicle, while still maintaining a high degree of parts commonality with the M151 and M151A1 vehicles.

Ford succeeded in designing a new rear suspension that corrected the handling, and maintaining extensive parts commonalty. In lieu of the swing arm type rear suspension previously used, the improved design featured semi trailing arm type rear suspension. With this suspension, to the driver, the MUTT handled more like a conventional car during turns, even when unloaded. This, it was hoped, would reduce the number of roll-over accidents by providing the drivers with an intuitive warning. Initially designated M151A1E1, it was soon given the Limited Production (LP) type classification of M151A2. Procurement of 24,503 of the redesigned vehicles was approved on 25 July 1969.

In addition to the revamped suspension, the M151A2 included many other improvements. Among these were deep-dish steering wheels, passenger grab handle, electric windshield wipers, larger, so-called "composite" marker and taillights, and a mechanical (as opposed to the earlier electrical) fuel pump. At the rear of the canvas top, a one-piece clear plastic rear window was added.

During the post-Vietnam era, the M151A2 and its variants remained in production, then by the General Products Division of Jeep Corporation, and its successor, AM General.

M151A1 vehicles, like this one, featured improved suspension systems. As the M151A1 was coming into service, so was a new machine gun, the M60. This combination was in use in Vietnam. (Kevin Emdee collection)

Work continued to improve the stability of the vehicles, resulting in the M151A1E1, shown here. This vehicle replaced the swing arm rear suspension with a trailing arm suspension. Once standardized, vehicles in this configuration were designated M151A2. Few M151A2s saw service in Vietnam. (Kevin Emdee collection)

The M37

However, the lack of commonality of parts throughout the tactical vehicle fleet during WWII had resulted in vast numbers of spare parts being required, and occasional bottlenecks in supply. Also, the lessons learned in that war brought about notable an ambulance, a command truck, a telephone line maintenance truck, and contact maintenance variant, a panel truck, two types of crash trucks and a handful of dump and wrecker truck variants.

The Dodge Model T-245 inline 6-cylinder 230.2 cubic inch engine powered the new trucks, with a Borg and Beck model 11828 clutch connecting it to a four-speed transmission and separate two-speed transfer case.

Two pilot models of the new vehicle were completed in December 1950—a scant (at least in Detroit terms) 19 months after the contract for it had been signed. The first production truck was completed in January 1951, and the vehicles continued to roll from the assembly lines established in Dodge's Route Van plant until July 1954. At that time the tooling for the vehicles was placed into storage—with considerable uncertainty as to whether it would ever be used again. This trepidation proved to be unfounded, for in less than three years the tooling was retrieved, and production resumed, with the first of the new series being completed in February 1958.

This second group of trucks differed in detail from the initial production, and the cargo variant was designated the M37B1. They can be most quickly distinguished from the first series by the spare tire's relocation to outboard of the driver's door, utilizing the mounting hardware initially developed for the M43 ambulance. Improvements were also made to the electrical system, as well as an improved transmission. The M37B1 remained in production until 1968, and was widely used in Vietnam as a utility vehicle.

The M37 was developed to replace the WWII-era Dodge WC-51/52 series of the 3/4-ton 4x4 trucks. The WWII-style combat wheels and other details of this example indicate that it is one of the three pre-production pilot models with a standard winch built by Dodge between December 1949 and May 31, 1950. (TACOM LCMC History Office.)

Plans for the post WWII U.S. tactical vehicle fleet included from the outset a 3/4-ton 4x4 truck. This "pickup truck"-sized vehicle had proven indispensable during WWII, and the G-502 (WC) series of trucks had been well-liked by the using forces—and naturally a good seller for Dodge.

advances across the entire range of U.S. tactical vehicles.

Dodge was naturally interested in maintaining their relationship with the Army, and accordingly created a truck that even today incorporates the best of both worlds—the proven components of the WWII vehicle as well as the advances of the "M-series" standardization.

Heavier gauge metal, including metal doors with roll-up windows, 24-volt sealed waterproof electrical system and a synchronized transmission were notable improvements over the earlier WC vehicles. The new trucks were also slightly narrower and had a slightly lower height than the earlier trucks as well.

Dubbed the G-741 series, the cornerstone of which was the M-37 cargo truck, in U.S. use the family included

This production M37 is equipped with an .30-caliber antiaircraft machine gun. The gun mount, M24A1, seen installed in this March 1951 Detroit Arsenal photograph, required a modification of the spare tire mounting, which on the M37 was located in the bed. (TACOM LCMC History Office.)

The M37 family of vehicles could be armed by using the M24A3 mount described later in this book. However, to install the mount the spare tire carrier of the M37 had to be removed and reinstalled, as shown here. The next generation of Dodge, the M37B1, solved this problem. (NARA)

Introduced in 1958, the M37B1 is readily distinguished by its door-mounted spare tire carrier. Almost 44,000 of this model were built, with less than 5% of the total being equipped with a winch. In contrast, almost 25% of the basic M37 trucks were winch-equipped. (Vintage Power Wagon collection)

The "Quad"

Seeking defense against enemy aircraft during WWII, the US Army sought a multiple machine gun air defense system. After various attempts, including mounting aircraft turrets on ground vehicles, the army settled on the M33 twin .50-caliber mount from W. L. Maxson. This soon gave way to the M45 from the same company, which upped the number of heavy machine guns to four. The turret was operated electrically; the power was supplied by batteries, which in turn were charged by a 300-watt generator, itself driven by a Briggs and Stratton gasoline engine.

The M45 could be installed on a number of carriages, including the halftrack, which became the M16 or M17 Multiple Gun Motor Carriage; the large M17 trailer, becoming the M51; or the small M20 trailer, becoming the M55.

These weapons were again used in Korea, and continued to be issued for many years thereafter. The mounts, however, were modified. The original Briggs and Stratton engines, well beyond their serviceable age, were replaced with new "Military Standard" engines. Standard rectangular ammunition boxes supplanted the tombstone-shaped ammo boxes common during WWII.

Each of the Air Defense Artillery Battalions that were deployed to Vietnam took a battery of M55 quad mounts, initially mounted in a 2¹/₂-ton truck using a Loading Aid Kit 7069664.

Battery G, 65th Artillery and Battery D, 71st Artillery were the first of the type to deploy to Vietnam, arriving in November 1966. Battery E, 41st Artillery, arriving in 11 March 1967, and Battery G, 55th Artillery, which was deployed 26 February 1968, joined them.

Although the Quad mounts were of little use against jet aircraft, they remained a formidable weapon against ground targets, both in perimeter defense as well as convoy protection. Charles E. Kirkpatrick reported in the spring 1989 issue of *Vietnam* magazine that these units fired 10 million rounds of .50-caliber machine gun ammunition while in Vietnam.

Above right: Some Vietnam gun trucks were armed with the .50-caliber multiple gun carriages M55, a powered, quadruple .50-caliber machine gun mount developed and put into service during World War II. Seen here is an example of a prototype version of that gun mount, designated the Multiple Caliber .50 Machine Gun Trailer Mount T81E1, comprising a Multiple Caliber .50 Machine Gun Mount M45E1 installed on a Mount Trailer T45E1. (Rock Island Arsenal Museum) **Below left:** With the aid of a winch in the truck bed, and issued ramps, Air Defense Artillery units were expected to manhandle the M55 mount in and out of the issued trucks. Here Japanese troops are trained in the use of this equipment at Fort Bliss. **Below right:** Starting in 1966, four batteries of the quadruple (quad) .50-caliber Multiple Gun Carriage M55 (G-65th, D-71st, E-41st and G-55th Artillery) were deployed to Vietnam. The four guns had an impressive combined rate of fire of up to 2,000 rounds per minute. (Air Defense Artillery Museum, both)

The .50-caliber multiple gun carriage M55 also went under such designations as the Quad .50-caliber Machine Gun M55 Mount, Trailer. This example, comprising an M45 mount on an M20 trailer, was photographed at the Rock Island Arsenal in Illinois in July 1967. (Rock Island Arsenal Museum)

At the rear of the mount is the battery charger, with the box-shaped gasoline tank to its right and exhaust to its left. By this time, the army had replaced the original power chargers on these mounts with military-standard units. (Rock Island Arsenal Museum)

TIRE PRESSURE 50 LBS.

The four Browning M2 HB .50-caliber machine guns of the M55 multiple gun carriage are at full elevation of 90 degrees. By the time of the Vietnam War, the U.S. Army had replaced the original 89 pound, 200 round tombstone-shaped M2 ammunition chests with plain 100 round ammo boxes, as seen here. (Rock Island Arsenal Museum)

To the lower left of the battery charger is the storage battery. Stenciled on the inside surface of the right trunnion is a caution notice for the gunner to check the interrupter for the correct setting before firing the guns. (Rock Island Arsenal Museum)

In another photo taken at Rock Island Arsenal in July 1967, the quad guns are traversed to the right, with the lunette (or towing eye) and drawbar in the foreground. Below the lower machine gun is the left trunnion sector; pinions drove the sectors to elevate the guns. (Rock Island Arsenal Museum)

A shield formed of thin armor provided some protection for the gunner. Attached to the trunnions above the upper machine guns are brackets holding a lateral rod to which the gun sight was attached. Various types of sights were available for use on this mount. (Rock Island Arsenal Museum)

TIRE PRESSURE
50 LBS.

Lifting rings were attached to the frame of the mount, seen here adjacent to the barrels of the lower machine guns. The gun mount could be traversed at a rate of up to 60 degrees per second and could be elevated at a rate of up to 60 degrees per second. The elevation limits were -10 to +90 degrees. (Rock Island Arsenal Museum)

Armored shields of different widths were mounted on hinges at the top of the lower armored shield. The gap between these two small shields provided clearance for the gun sight. The gun sight tilted upward and downward in unison with the elevating or depressing of the guns. (Rock Island Arsenal Museum)

On the corners of the trailer were ratchet-type jacks for the purpose of leveling the trailer and the gun mount. A jack handle is inserted in the left right rear jack. On the rear of the trailer are a manufacturer's and data plate, a taillight assembly, and a reflector. (Rock Island Arsenal Museum)

The quad .50-caliber machine guns had a minimum elevation of -10 degrees. A single M2 HB .50-caliber machine gun packed a punch sufficient to disable or destroy aircraft, soft skin vehicles, lightly armored vehicles, or massed infantry. The combination of four guns provided truly impressive firepower. (Rock Island Arsenal Museum)

The Birth of the Gun Truck

Most military organizations are noteworthy for their uniformity. Clothing, weapons, vehicles—all the same within a nation or branch, or certainly within a unit. The absence of this uniformity is perhaps one of the reasons that the gun trucks used in Vietnam have achieved something of celebrity status. The bravery and innovation of the crews of these vehicles further enhanced their mystique.

Faced with ongoing ambush attacks, particularly along Highway 19 (the same route along which French Mobile Group 100 had been annihilated a decade earlier), the truckers resorted to first protecting themselves with armor, and almost immediately arming themselves as well. Few people today recognize the enormity of the task these truckers faced.

Most readers are familiar, and duly impressed with, the feats accomplished by the truckers of the Red Ball Express during WWII. This unit, operating almost 6,000 trucks, moved an average of 12,500 tons of supplies a day, with a total movement of about a half-million tons of cargo during its 25 August through 16 November, 1944 duration of operation. The route was 650-750 miles long. This undertaking was largely unmolested by enemy operations.

In Vietnam twenty-two years later, the Eighth Transportation Group faced a similar situation—supply huge quantities of supplies to troops operating in the field, far from ports. During the quarter-year ending July 1968, Eighth Transportation Group vehicles traveled 5,557,000 miles and carried 278,000 tons of cargo, 1,200,000 gallons of POL, and 22,988 passengers—despite frequent ambushes. And that quarter year was but a sliver of a multi-year operation.

The Viet Cong were determined to close Route 19, or even better, annihilate an American unit. The U.S. commanders were equally determined to keep Route 19 open.

By late 1967, Viet Cong ambushes of U.S. convoys in Vietnam had reached an alarming level—the breaking point being reached on September 2, when a company-sized VC force attacked thirty-nine Eighth Transportation Group trucks in convoy near An Keh. Thirty-four of the vehicles were damaged or destroyed—worse, 17 GIs were killed in action and 13 more wounded.

Up to this point, the armament for these convoys, which at times exceeded 100 vehicles, consisted of the soldier's rifles and the occasional vehicle equipped with an anti-aircraft machine gun. Following this event, efforts began in earnest to provide convoy protection. Though prior officers had made some attempts in this effort, Col. Bellino, commanding officer of the Eighth Transportation Group, truly became a champion of this concept, allocating the resources to allow his men to experiment in this area and promoting the idea to his superiors.

The earliest attempts at "hardening" a cargo truck consisted of parallel walls of PSP plating, with sandbag filler between them. Adding the rain typical of Southeast Asia to this mix resulted in a very overloaded 2½-ton truck. So, a single layer wall of prefabricated 36" x 48" armor plates replaced the PSP and sandbags, resulting in a truck operating closer to its design specifications. Armament at first was small arms, then later M60 7.62 machine guns were added and then .50-caliber M2 HB machine guns began to be added. The TOE for transportation units did not provide for sufficient weapons, so the truckers traded for weapons, acquired them on the black market, or pilfered weapons from damaged equipment they were hauling. By cannibalizing these weapons, they were able to build their own working arsenal.

Although the Quad .50-caliber machine gun batteries of the air defense artillery units were supplied with WWII-era M45 quadruple .50-caliber machine gun mounts, including truck mounts, they lacked armor protection. Improvised armor began to be installed, often consisting of armor plate on the floor, behind the seat, and slipped into the window channel in the door.

The quad-fifty mount, though effective, was issued only to four batteries, and thus was in short supply. The armored box, fabricated by transportation and in some cases engineer units, was far more common. However, the roads in Vietnam were in large part in terrible condition, and the pounding could shake loose the single wall of plating hung on the outside of the bed. Adding a spaced second wall of armor plate inside not only offered greater protection, but the two walls were connected to support each other. GIs then began to fill the gap between the walls with spare tires, ammo, and sandbags—resulting, once again, in an overloaded vehicle.

While the 2½-ton gun trucks had proven the soundness of the concept, the heavy armor plating, weaponry, and the vast stores of ammo amassed by the truckers overburdened the chassis. In at least one case a frame failure resulted! Accordingly, the men turned to the 5-ton trucks that made up the bulk of the transportation units' vehicles. These trucks had not been used from the outset in hopes of preserving as much of the units' transport ability as possible. However, it was ultimately decided that more tonnage could be moved with fewer escorted vehicles than more trucks could move with inadequate escort. At this point, 5-ton trucks began to largely replace the 2½-ton trucks in the convoy escort role.

Few 5-ton trucks were built with single wall armor plating, with the double wall armor box being much preferred.

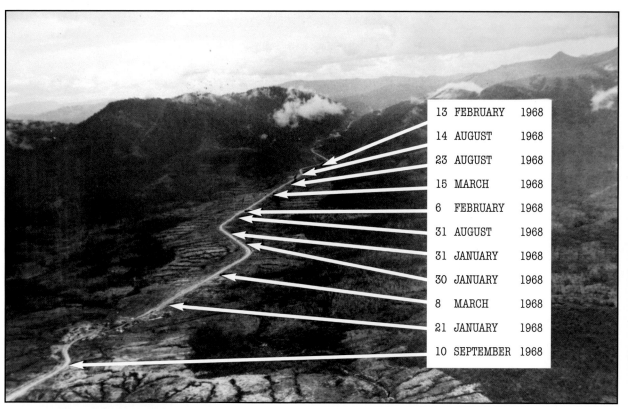

13	FEBRUARY	1968
14	AUGUST	1968
23	AUGUST	1968
15	MARCH	1968
6	FEBRUARY	1968
31	AUGUST	1968
31	JANUARY	1968
30	JANUARY	1968
8	MARCH	1968
21	JANUARY	1968
10	SEPTEMBER	1968

This wartime illustration indicates the frequency of ambushes on one stretch of road in Vietnam during 1968. These unrelenting attacks against meagerly armed truckers brought about the creation of the gun truck.

This M54 with an experimental armor package was photographed in December 1965 at the army's Tank-Automotive Center near Detroit. While 650 of these Army kits would be produced, none would be used in Vietnam until mid-1967. (Jeff Houghton collection)

The persistent shortage of armor plate in 1968 meant that the truckers looked for an alternative. They found this in mounting an M113 Armored Personnel Carrier on a 5-ton truck. Even though the APC was stripped of its drive train to reduce weight, the resultant trucks were very top-heavy and difficult to drive. With the APC mounted in the bed facing forward the concentration of weight on the rear axles compounded the handling problems. Consequently, some units mounted the APCs backwards in order to better distribute the load. Despite this improvement, this style of truck remained unwieldy. Combining this deficiency with the reduced field of vision and limited number of weapon mounting points, the APC-carrying gun trucks began to be phased out in 1970.

The armored box-type of trucks went through a variety of weapons upgrades. M2 HB .50-caliber machine guns were mounted in single or twin mounts, and some trucks were armed with one, two, three, or even four GE XM134 7.62mm Miniguns. Though sensitive to dust and prone to jamming, the Miniguns could lay down an astonishing amount of firepower. The weapons had a selectable rate of fire, up to 6,000 rounds per minute, though the truckers usually fired at a rate slightly slower than the maximum.

The trucks themselves were M54A1, M54A2 and M54A2C 6x6s. The transportation units preferred the Mack ENDT-673-powered M54A1 for their transport work, feeling the engine had greater lugging ability than did the Multifuel. In the early days during operations in some mountain areas convoy speeds dropped as low as 4 MPH regularly, making torque a high priority. With the gun trucks drawn from the task trucks of the units, it was natural then that many of these were also Mack-powered vehicles.

A few M37 3/4-ton cargo trucks were hardened and armed in 1967 as well—but the overloading characteristic of the 2 1/2-ton gun trucks was even more compounded in the case of the Dodge. Several M151 MUTT vehicles were equipped with machine guns, and in some cases armor plating as well. This naturally and grossly overloaded the small vehicle, and no doubt made the handling of the vehicle "interesting."

Improved strategies and to a greater extent the effectiveness of the gun truck crews, resulted in diminishing convoy losses. As the U.S. forces began their drawdown most of the gun trucks were transferred to the Republic of Vietnam, where almost invariably the armored boxes were removed and the trucks reverted to their cargo roles.

The kit included a modest amount of plating for the windshield. This was subsequently augmented with bullet-resistant glass. (Jeff Houghton collection)

Weapons

Gun truck crews, as stated earlier, were somewhat resourceful when it came to arming themselves and their vehicles. Weaponry ranged from pistols to antiaircraft armament dating back to WWII on up to the most modern Miniguns, modern-day Gatling guns with withering firepower.

Since WWII, U.S. military tactical vehicles have been designed to accept modification kits to permit the mounting of antiaircraft machine guns. Of course, these weapons can also be used against ground targets. In the case of the 2 1/2-ton and 5-ton 6x6 trucks, this kit consisted primarily of three support legs, which were attached to the cab via clamps, two on the rear of the cab and one just forward of the passenger's door, a ring mount, which those legs supported,

and the pintle and cradle assembly which actually secured the weapon to the ring. The ring was supported over the truck cab, and the passenger's seat back folded forward on the seat bottom. The back of the seat, which was a diamond-tread anti-slip surface, became the gunner's platform. The ring itself could be one of several styles of the M49, an M66, or an M66C, with the supporting structure varying widely depending on the vehicle application. The M49 was the ring mount most commonly used on 6x6 trucks during the Vietnam era. A later variation, occasionally seen on vehicles in Vietnam, was the M49A1. The ring mount featured a backrest for the gunner to lean against. This backrest rotated in conjunction with the pintle and was intended to provide a means of better tracking target aircraft. The ring itself and supports are identical to that used in the standard mounts. Still later, the M66 was introduced. The M66 was roller bearing-equipped, allowing easier and more rapid traverse of the weapon and gunner.

Above: The M49 ring used carriage assembly D40721 that allowed the weapon to be traversed around the gunner, as well as being pivoted on its pintle. **Below:** Initially, the ammunition supply was supported by tray D40731, but the tray D90078, which is shown here, later superseded it. **Bottom:** The pintle assembly used was the D40733, which provided for the weapon to be elevated 80 degrees, or depressed 20 degrees. (Rock Island Arsenal Museum, all)

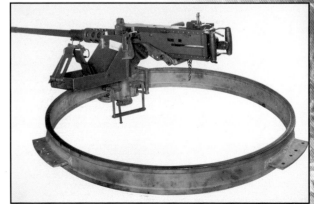

Some gun trucks were armed with 7.62mm XM134 GE Miniguns originally used as helicopter armament. The electrically operated modern-day six-barrel Gatling gun could fire up to 4,000 rounds per minute, an intimidating and lethal response to convoy attacks. A rate of fire of 6,000 rounds per minute was possible, but the lower rate was considered more prudent to avoid overheating. (NARA)

This device is the bracket developed to install a Multiple Caliber .50 Machine Gun Mount M45 on a gun truck. The bracket was attached to the floor of the truck body, and the gun mount was removed from the M20 trailer and fastened to the bracket.

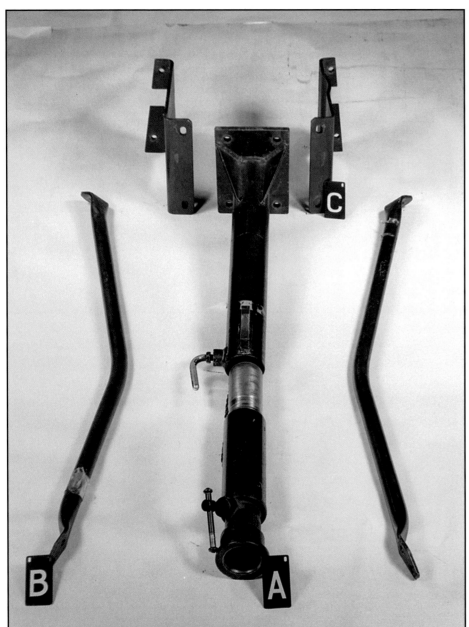

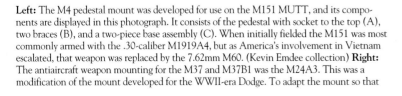

Left: The M4 pedestal mount was developed for use on the M151 MUTT, and its components are displayed in this photograph. It consists of the pedestal with socket to the top (A), two braces (B), and a two-piece base assembly (C). When initially fielded the M151 was most commonly armed with the .30-caliber M1919A4, but as America's involvement in Vietnam escalated, that weapon was replaced by the 7.62mm M60. (Kevin Emdee collection) **Right:** The antiaircraft weapon mounting for the M37 and M37B1 was the M24A3. This was a modification of the mount developed for the WWII-era Dodge. To adapt the mount so that it would reach the edge of the M37 and M42 beds, extensions were added to the ends of the mount, This mount was intended to be installed across the front of the bed, with the "post" on the side of the mount facing the bed, and is supported by the ends as well as bolting to the floor of the vehicle. (Rock Island Arsenal Museum)

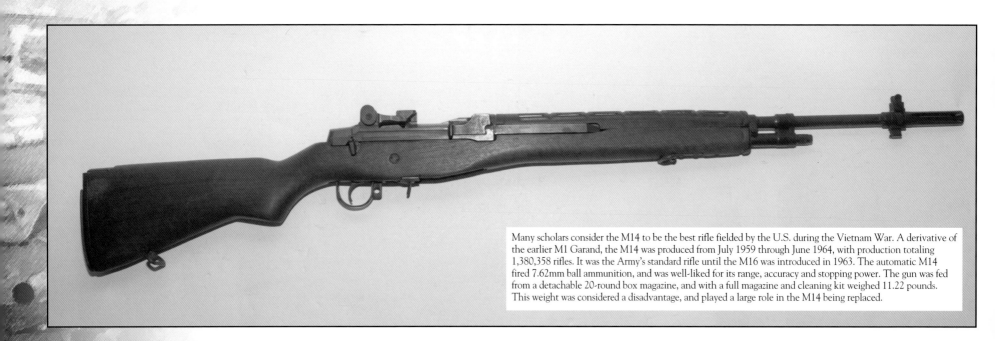

Many scholars consider the M14 to be the best rifle fielded by the U.S. during the Vietnam War. A derivative of the earlier M1 Garand, the M14 was produced from July 1959 through June 1964, with production totaling 1,380,358 rifles. It was the Army's standard rifle until the M16 was introduced in 1963. The automatic M14 fired 7.62mm ball ammunition, and was well-liked for its range, accuracy and stopping power. The gun was fed from a detachable 20-round box magazine, and with a full magazine and cleaning kit weighed 11.22 pounds. This weight was considered a disadvantage, and played a large role in the M14 being replaced.

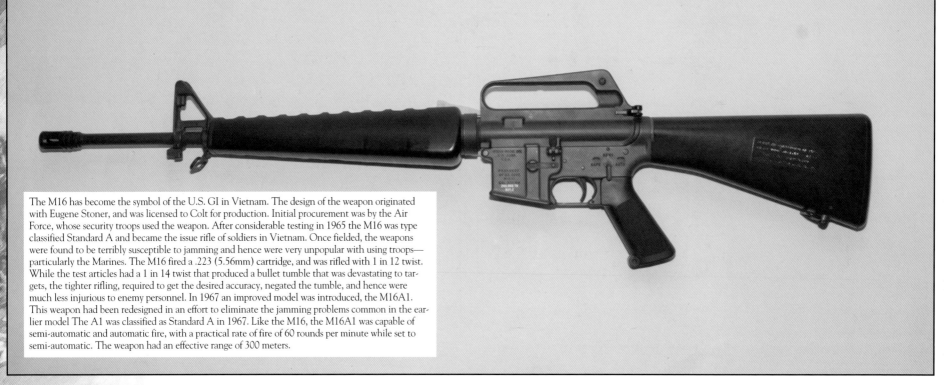

The M16 has become the symbol of the U.S. GI in Vietnam. The design of the weapon originated with Eugene Stoner, and was licensed to Colt for production. Initial procurement was by the Air Force, whose security troops used the weapon. After considerable testing in 1965 the M16 was type classified Standard A and became the issue rifle of soldiers in Vietnam. Once fielded, the weapons were found to be terribly susceptible to jamming and hence were very unpopular with using troops—particularly the Marines. The M16 fired a .223 (5.56mm) cartridge, and was rifled with 1 in 12 twist. While the test articles had a 1 in 14 twist that produced a bullet tumble that was devastating to targets, the tighter rifling, required to get the desired accuracy, negated the tumble, and hence were much less injurious to enemy personnel. In 1967 an improved model was introduced, the M16A1. This weapon had been redesigned in an effort to eliminate the jamming problems common in the earlier model The A1 was classified as Standard A in 1967. Like the M16, the M16A1 was capable of semi-automatic and automatic fire, with a practical rate of fire of 60 rounds per minute while set to semi-automatic. The weapon had an effective range of 300 meters.

Top: The M203 grenade launcher, introduced in 1969, was designed as an attachment for the M16A1 rifle, negating the need for the separate M79 grenade launcher. This was a semi-permanent attachment performed by armorers. The 12-inch long rifled grenade launcher fired 40mm ammunition that came in a variety of forms, including smoke, high explosive, illuminating, High Explosive Dual Purpose, and a variety of others. The weapon had a range of about 400 yards, but was not notable for its accuracy. **Above:** Frequently referred to as "Thumper" or "Blooper," the M79 was first issued in 1961. Rather resembling a very large, short shotgun, the M79 was a single shot, break barrel weapon firing a 40mm grenade. It was intended for use was for uses between the ranges of hand-thrown grenades and mortars. Two M79-armed grenadiers served in each rifle company. These men, in addition to being armed with the M79, also carried M1911A1 pistols.

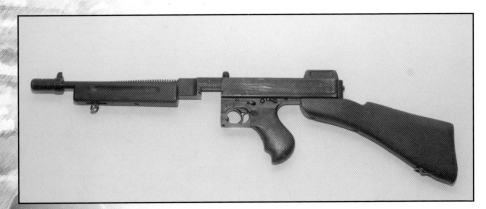

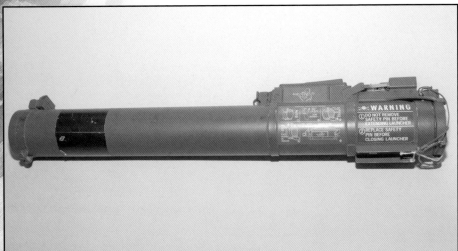

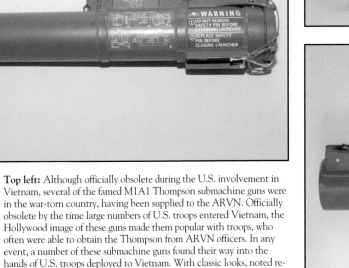

Top left: Although officially obsolete during the U.S. involvement in Vietnam, several of the famed M1A1 Thompson submachine guns were in the war-torn country, having been supplied to the ARVN. Officially obsolete by the time large numbers of U.S. troops entered Vietnam, the Hollywood image of these guns made them popular with troops, who often were able to obtain the Thompson from ARVN officers. In any event, a number of these submachine guns found their way into the hands of U.S. troops deployed to Vietnam. With classic looks, noted reliability, and easy access to ammunition (the Thompson fired the same ammunition as the M1911A1 pistol) many soldiers added the "Tommy gun" to their personal arsenal. **Top right:** The M3A1 .45 caliber submachine gun, popularly known as the "grease gun," was developed during WWII as a low-cost, easily produced replacement for the Thompson submachine gun. During WWII the Guide Lamp Division of General Motors, the originator of the design, produced the gun, and during the Korean War Ithaca manufactured additional copies. By the time the U.S. entered Vietnam, the grease gun was no longer in widespread use by the army, however the weapon did continue to be used by ARVN troops and Marine armored crewmen. Black market deals, trades and pilfering of destroyed vehicles resulted in some gun truck crewmen equipping themselves with the venerable weapon. **Above left:** The M72 Light Antitank Weapon (LAW) was originally developed as an infantry weapon to combat the hordes of Soviet tanks that threatened Europe during the Cold War. It was classified as Standard in March 1961.

Because the Communists made only limited use of armored vehicles in Vietnam, the LAW was most frequently used for bunker busting. The disposable lightweight weapon weighed about 5 pounds. With the end covers shut, the weapon was watertight. To fire the weapon the end covers were opened and an inner tube slid out. This action cocked the launcher. Pulling a trigger fired the 1-kg rocket, which had a 66mm high explosive antitank warhead, after which the launcher was discarded. It had an effective range of 300 meters. **Above right:** The first M72 LAWs used in Vietnam were notable for their failure to fire. Immediate efforts were made to counter this, and the M72A1 shown here was fielded in response. Even this was not sufficient and a third variant, the M72A2, was also developed and used in Vietnam, further enhancing reliability. (Rock Island Arsenal Museum, both)

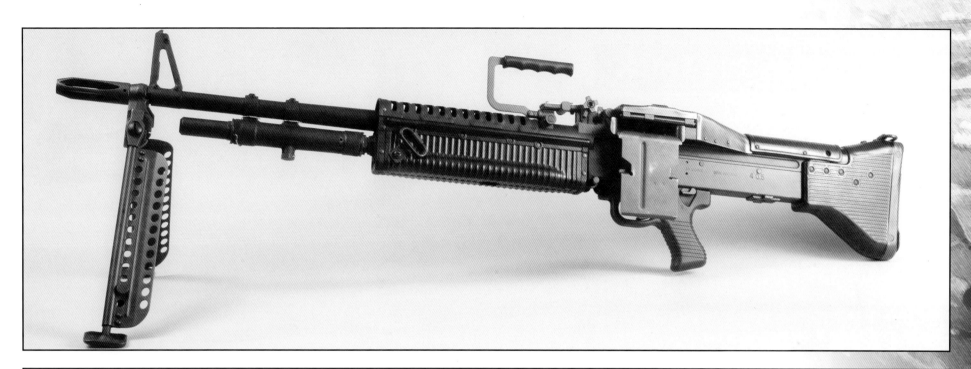

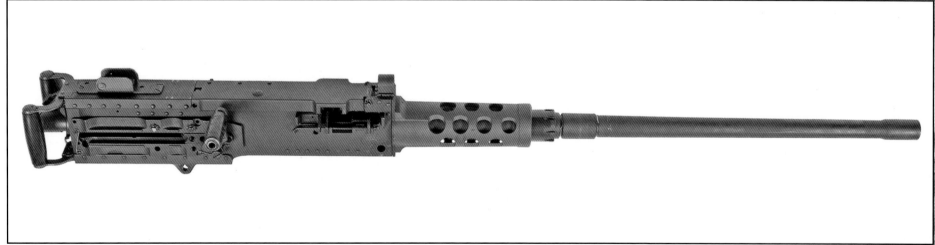

Top: The M60 was the primary squad weapon in the jungles of Vietnam. With a maximum rate of fire of 550 rounds per minute, the gun had a voracious appetite for ammunition. The 7.62-mm M60 was used as a door mount weapon on helicopters, as well as a mounted weapon on various tactical vehicles, but its primary use was as a squad weapon. For this use the weapon was equipped with a bipod. Infantrymen in Vietnam complained about the weapons 23-pound weight and tendency to jam, particularly when exposed to dust. The gun was supplied with an asbestos glove to be used when changing barrels that were hot from sustained fire. **Above:** The Browning .50-caliber machine gun was the Army's primary heavy machine gun in Vietnam, as it had been in WWII and Korea, and continues to be to this day. The .50-caliber design dates to 1921, considered by many to be John Browning's finest design, was originally water-cooled. In 1932 an air-cooled version for ground use was developed. This version had a barrel with thicker walls to withstand the heating. This resulted in the barrel being denoted heavy, and the weapon we know today as the .50-caliber machine gun is properly the M2 HB (Heavy Barrel) machine gun. Though widely used on various vehicle mountings, the M2 was designed as an infantry weapon, and was widely used as such. In this use the 84-lb machine gun was mounted on a M2 tripod, which weighed a further 44 pounds. The gun has a cyclic rate of fire of 450-550 rounds per minute, but rates of less than 400 rounds per minute are typical in use. The maximum range of the weapon is 4.5 miles, although its effective range is about 1.2 miles. (US Army, both)

The 2½-ton Gun Truck

When U.S. troops began deploying in numbers to Vietnam, the standard army cargo truck was the gasoline-powered M35, seen here on the beach. Convoys of these vehicles, and their larger 5-ton siblings, stretched across the country. Their lack of defense, particularly when laden with cargo rather than armed infantry, made these convoys highly vulnerable to enemy attack. (NARA)

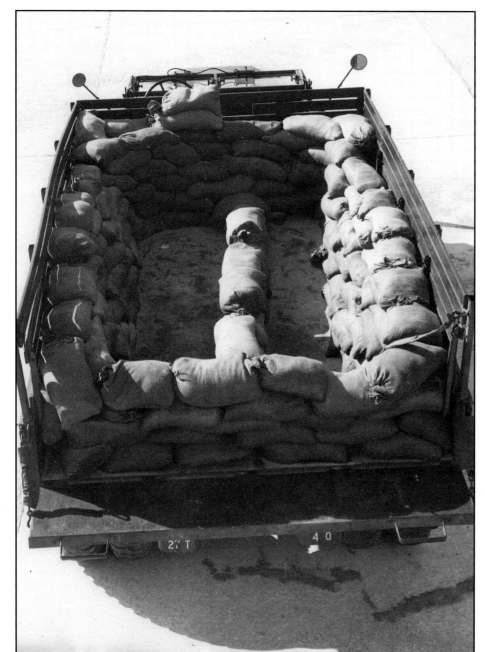

Left: An early expedient to harden trucks for convoy security was to stack sandbags in the cargo bed, to form a sort of mobile firing pit, as in this example from the 27th Transportation Battalion. When rained on, however, the sandbags became very heavy. **Top right:** The driver of "Old Reliable Express" has improvised a moderate level of protection by lashing what appears to be a number of lightweight armor panels to the door. The panels look similar to the type mounted in UH-1 helicopters to protect pilots from small arms fire from the sides. (NARA) **Above right:** "The Gamblers," a 1967 Kaiser-Jeep M35A2 of the 27th Transportation Battalion, 8th Transportation Group, was hardened with armored kit panels with firing apertures and was fitted two M60 machine guns and a ring mount for an M2 .50-caliber machine gun.

A quad .50-caliber machine gun mount on a gun truck is observed close-up. The thickness, or rather the thinness, of the low armor on the sides of the cargo bed is visible. The muzzles of the machine gun barrels are fitted with conical-shaped flash suppressors. (Air Defense Artillery Museum)

This undocumented photo from the Patton Museum shows two early quad .50-caliber gun mounts from one of the four deployed quad-50 units. It is an early image ('67-'68) since the truck still retains the loading rails (one is visible at the bottom of the armor plate) that it was deployed with. Behind the cab is the winch and cable group used to load and unload the M55 mount from the bed of the 1967 Kaiser-Jeep M35A2 truck. Vertical mounts for the M60 MGs are also visible. The armor plating and improvised gun shields were added "in-country". (Patton Museum)

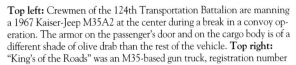

Top left: Crewmen of the 124th Transportation Battalion are manning a 1967 Kaiser-Jeep M35A2 at the center during a break in a convoy operation. The armor on the passenger's door and on the cargo body is of a different shade of olive drab than the rest of the vehicle. **Top right:** "King's of the Roads" was an M35-based gun truck, registration number 4L3451, that served with the 541th Transportation Company, 12th Transportation Battalion, 8th Transportation Group. Note how the rear of the passenger's door armor angles out. **Above left:** Many gun trucks owed their names to popular songs, such as this quad-.50-caliber-armed example named after the opening words of The Crazy World of Arthur Brown's 1968 hit, "Fire." Wing-armor plates were added to the gun mount as a field modification.

Note the M60 machine gun, winch and loading rails. (Air Defense Artillery Museum)
Above right: Above right: "Duce is Wild," assigned to the 541st Transportation Company of the 124th Transportation Battalion, was a 1967 Kaiser-Jeep M35A2. The .50-caliber machine gun has an ACAV-style shield and twin mounted M60 MGs. "The Tennesseean" was painted on the rear of the vehicle.

A well-travelled quad-.50 crew, three of whom are wearing flak vests, are serving their mount atop a 2½-ton truck alongside a road probably in the Central Highlands or the DMZ. The gunner is in the center position between the pairs of guns, and is somewhat protected by the armored shell in front of him. Two control handles to his front are used to electrically traverse, elevate, and fire the guns. The left loader is pointing something of interest out to the gunner. The Squad Leader oversees from behind. The right loader is moving away. Note that this particular unit does not have any additional armor plating along the sides of the truck or the gun shields (sometimes called 'bat-wings') attached to the M55 mount. The loading rails are still present with this unit. Flash suppressors are not present on the .50's. (Air Defense Artillery Museum)

Top left: Another Kaiser-Jeep M35A2 gun truck of the 27th Transportation Battalion was "Nancy," armed with a quad .50-caliber M55 mount. The truck was photographed in the latter part of 1968. Eyes and a mouth with sharp teeth were painted on the shield of the gun mount. **Top right:** "Nancy" is viewed from the left side as its crew enjoys a break from action. This color photo reveals that the markings applied by the crew on the gun mount shield and the armor plate at the front of the cargo bed were in yellow and red paint. **Above left:** This Quad-.50 gun truck belongs to D-Battery 71st Artillery and was stationed in III CORP / IIFFV, circa 1967. It is fully armor plated and has a makeshift canopy to keep the sun off the crew and equipment. Note the road wheels for the M55 mount are visible on the rear

tailgate. (Robert Rohrs, D-71st, NDQSA Archives) **Above right:** A quad-.50 M55 gun truck of G-Battery 65th Artillery shows the effect of a 1966-67 enemy rocket or mortar attack. The mount is still in the bed of the truck and minimal armor plating is visible. Note the finger writing in dust on the left, "SHORT 33 days" alluding to a crewmember's time left in his tour. (Alan Wentworth, D-1/44 Arty, NDQSA Archives)

Top left: LTC McIntyre, commander of 5th Battalion, 2nd Artillery, stands at the rear of a quad.-50 gun truck of the attached D-Battery 71st Artillery. Stenciled in yellow on that mount is "Judge / Jury / Executioner." The names and mottos painted on gun mounts were usually a separate item from the name of the gun truck. **Above left:** Gun trucks occasionally came to grief, such as this 1966 or 1967 Kaiser-Jeep M35A2, which evidently detonated a mine, destroying the front end. Worthy of notice are the curved cutouts on the bottom of the side armor, to accommodate the tandem tires. (Air Defense Artillery Museum, both) **Above right:** Although "Bad Dream" is emblazoned on the rear of this 1967 Kaiser-Jeep M35A2 gun truck of the 541st Transportation Company, 124th Transportation Battalion, 8th Transportation Group, its real name, painted on the side armor, was "Assassins."

In 1971, Specialist Don Long of G-65th Artillery (MG) sits in the gunner's seat of their M55 gun mount. There is no artwork on this particular gun's shield. Note the spent .50-caliber shell casings littering the floor of the truck indicating that it was recently fired. The M18 reflex sight is not present and one of the barrels is missing a flash suppressor. The winch and cable assembly for loading and unloading the mount from the truck is visible in the foreground, as well as the loading ramps visible on the sides of the trucks. Behind him is another G-65th quad on a 2.5-ton truck. (Lynn Burgess, G-65th, NDQSA Archives)

Two G-Battery, 65th ADA Quads, mounted in a pair of 1966 Kaiser-Jeep M35A2 trucks, await orders at a firebase in I Corps Tactical Zone. LD-465 naturally aspirated Multifuel engines, the exhaust outlets for which are visible above the rear tandems, power the trucks. The trucks exhibit the low-mount headlights, a feature that later gave way to the use of the high mount headlights, previously found only on winch-equipped 2½-ton trucks, on all of the 6x6s of this weight class. (NARA)

Top left: Gun trucks were also used as road security for engineer projects during the day. Here is another colorful 2 ½-ton gun truck from either an engineer's battalion or a transportation group circa 1971 on QL14 south of Pleiku. The Old English text reads "We Kill Dinks", "The Good Bad & Ugly" INC. with a cameo of the Zig-Zag man. (Dale Stiebritz, B-4/60th Arty, NDQSA Archives) **Top right:** This gun truck was stopped in its tracks by a command detonated mine just off the edge of a bridge somewhere in IICORP / IFFV, circa 1969. The gun box is taller than most and has two .50-caliber MGs with gun shields. A string of flowers lines the top of the gun box and a cartoon-like character is visible on the side, probably Thor from the B.C. comic strip. No text is visible. (Richard Iaeger, A-4/60th, NDQSA Archives) **Above left:** "Together Again," a Kaiser-Jeep M35A2 with the 27th Transportation Battalion, was outfitted with a type of multifaceted

armor called keyhole armor because when viewed from above it was keyhole shaped. The multiple angles were intended to deflect bullets. **Above right:** "Killer I Jr" was a 1966 Kaiser-Jeep M35A2, registration number 4J6566, assigned to the 88th Transportation Company. Several M60 machine guns on pintle mounts are visible above the armored box, which was mounted over the low, raised sides of the cargo body.

This gun truck, registration 4H6374, was based on a 1966 Kaiser-Jeep M35A2 powered by a LD-465 engine. Sandbags are piled on the running board and the windshield. Bolts for fastening the armor to the body were used sparingly and are light colored.

Gun trucks of the 572nd Transportation Company with Marines aboard prepare for a convoy mission in April 1968. The nearest vehicle is a 1966 Kaiser-Jeep M35A2. The second truck in line is "Kong Killer." In the background is a quad .50-caliber gun truck. Partially hidden in the background is a M42 Duster self-propelled antiaircraft gun, armed with two 40mm Bofors automatic guns. (NARA)

Down time: A soldier in the M151 is airing out his feet waiting for road clearance or escorts. A gun truck with the 19th Engineers is parked behind a scraper somewhere on QL19 west of Qui Nhon circa 1969-70. The Engineer unit is probably heading west, through the ambush-prone An Khe and Mang Yang passes to road building operations in the Central Highlands. A metal ladder is on the side of the gun box allowing the gunners to climb aboard quickly. Two M60 machine guns are pointed right and a .50-cal is pointing left. Spare tires also serve as protection from small arms fire. (Lynn Wood, E-41, NDQSA Archives)

The 5-ton Gun Truck

The rugged terrain, heavy use, and recoil of the quad-fifty proved to be a considerable burden on the 2$^1/_2$-ton 6x6 chassis. Soon ADA units began using the more robust 5-ton 6x6 as a platform for the M55 mount. A 1967 Kaiser-Jeep M54A2 formed the basis for this unmarked quad .50-caliber machine gun equipped gun truck, photographed in mid-September 1970. The M54A2 appeared similar to the M54A1 but had a vertical exhaust extending through the rear of the right fender. An extended steel platform sits in the bed of the truck and there is no side armor protection visible except for the driver's side door. (US Army Quartermaster Museum)

For the ADA units—and in fact all gun trucks—the truck is merely the carriage for the gun—and the weapon had to be serviceable as much as possible. In order to achieve this, the weapons—whether the mounting of the M55 of ADA units, or the "gun boxes" used by transportation units—were constructed such that they could be moved quickly from truck to truck. These two Battery E, 41st Artillery M55 Quad .50-cal gun mounts sit atop a fabricated steel platform that slid into the back of the 5-ton cargo truck. The platforms extended over the tailgate, giving the crew more foot and storage room, as well as having the added benefit of additional armor protection. At left in this 1971 image is E-54 "Freedom Fighters," featuring an American Flag on the gun-shield and the Korean and South Vietnamese flags on the "bat-wings." At right is E-53 "Young Crusaders," with "Whole Lotta Lead" across the back plate. (Dennis Ruth E-41st, NDQSA Archives)

At base camp in Tuy Hoa, 1971, two E-battery 41st ADA M55 Quad .50-cal gun mounts sit atop a fabricated steel platform that slides into the back of the M54A2 5-ton cargo truck. The platform extends over the tailgate giving the crew more foot and storage room. On the left is "American Breed" with an eagle on the gun-shield. The gun shields of "American Breed" are intricately formed, fitting snugly to the gunner's armored shell. On the right the loader's shields have the Confederate battle flag.

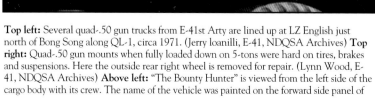

Top left: Several quad-.50 gun trucks from E-41st Arty are lined up at LZ English just north of Bong Song along QL-1, circa 1971. (Jerry Ioanilli, E-41, NDQSA Archives) **Top right:** Quad-.50 gun mounts when fully loaded down on 5-tons were hard on tires, brakes and suspensions. Here the outside rear right wheel is removed for repair. (Lynn Wood, E-41, NDQSA Archives) **Above left:** "The Bounty Hunter" is viewed from the left side of the cargo body with its crew. The name of the vehicle was painted on the forward side panel of the cargo body, as faintly seen in the preceding photograph, as well as on the machine gunner's armored shell. **Above right:** The quad mount of "The Bounty Hunter" is viewed close-up, showing details of the machine gun cradles and gun sight. The lettering of the name was red with scraggly white border. The truck's march-order number, 74, was stenciled on each 200 round ammunition box.

Sgt. Bill Simpson of G-Battery 65th Artillery (MG) sits next to his quad-.50 caliber gun mount and 5-ton truck in 1970 near Mai Loc. Stenciled across the gun shield is "PARIS PEACE TALKS," a wry homage to ongoing negotiations with the North Vietnamese. At the base of the gun shield is an unofficial insignia of Battery G, 65th, a red "G" with four blue .50-cals on a green background. There is little to no armor plating on this truck but for the front shell of the M55 mount. Many of G-65th guns were actually deployed to remote bases not accessible by truck. They were airlifted in, thus the need for armor plated trucks was not the same as other quad-50 batteries in other parts of Vietnam. Personal gear hangs from various locations and the driver appears to have upgraded plated body armor. Another crewman lays in the shade beneath the 1966 M54A2 reading a paper. (Lynn Burgess, G-65th, NDQSA Archives)

Top left: This quad .50 gun truck was part of G-Battery 55th Artillery, which based out of Chu Lai and was attached to the Americal Division, circa 1971. Again, as with most quad-.50 gun trucks it had minimal armor plating. The truck is towing a M105 trailer for ammo and supplies. (Joe Kotarba, G-55th, NDQSA Archives) **Top right:** "Young Crusaders," an E-41st Quad, rests between assignments. The slide-in insert supporting the M55 has clearly been sized precisely to fill a M54 bed to the end of the tailgate in the lowered position. (Jim

Feil, E-41st NDSQA Archives) **Above left:** Although Rock Island Arsenal developed precisely fitting bat-wing armor for the M55 mount, the Quad units in Vietnam had to rely on their own locally fabricated gun shields, an example of which is seen here. (NARA) **Above right:** "Backfield in Motion," a 5-ton truck with E Battery 41st Artillery, rounds a bend on QL-19. This truck, which has an unusual camouflage pattern on its gun shield, was later renamed "Freedom Fighters." (Lynn Wood, E-41st NDSQA Archives)

"The Hawk," an M54A2, probably was a 1967 Kaiser-Jeep example. The art on the early-type armor plates on the cargo body consists of a drawing of a hawk with "The Hawk" written across the middle. It had one .50-caliber machine gun and two M60s.

"Black Widow" of the 523rd Transportation Company, 54th Transportation Battalion, 8th Transportation Group, is an example of a gun truck based on the 5-ton M54A2C cargo truck. This model is identifiable by the air cleaner canister on the right fender and the engine exhaust pipe under the fender. "Black Widow" was armed with three pedestal-mounted M2 .50-caliber machine guns.

Top left: As viewed from behind a pedestal-mounted .50-caliber machine gun, a gun truck proceeds down a highway in the Republic of Vietnam. On the floor are tightly packed .50-caliber ammunition boxes, a stowage practice that was common in gun trucks. **Top right:** A crewman with one hand on the grip of an M2 .50-caliber machine gun talks into a radio handset. Radio communications were vital in gun truck operations, as the crews needed instant and accurate information on the convoy's situation to react quickly to attacks. (US Army Quartermaster Museum, both) **Above left:** Named after a 1968 album by the Mothers of Invention, "Uncle Meat" was a 1969 Kaiser-Jeep M54A2C. This vehicle was photographed with several different armament configurations. Here it has a twin .50-caliber machine gun mount in the rear and a single .50-caliber mount forward. **Above right:** "Uncle Meat," center, is viewed from the front, displaying markings for the 503rd Transportation Company, 124th Transportation Battalion, 8th Transportation Group. Various details on the front end of the vehicle were decorated with white paint.

Machine guns dismounted, "Uncle Meat" rests in an equipment yard at Fire Support Base Vandergrift, Quang Tri Province, during Lam Son 719, an offensive operation that lasted from February to April 1971. The tie-down hooks on the cargo bed are painted white.

Top left: "Cold Sweat" was an M54A2C-based gun truck that served successively with several companies of the 8th Transportation Group. The original vertical exhaust was converted to under the fender. The front of the hood was yellow. The air cleaner clamps were white.
Top right: The gun truck "The Justifier," based on a 1969 Kaiser-Jeep M54A2, served with the 57th Transportation Company. This truck had several different designs of "The Justifier" lettering over time. This version was white at one time and was red in this photograph.

Above left: "Born Loser" had early-style kit armor with precut firing ports. The vehicle, an M54A2C, saw action with the 585th Transportation Company, 27th Transportation Battalion. The white shape between "Born" and "Loser" are aces and eights playing cards. **Above right:** "Old Grand-Dad," of the U.S. Army's 515th Transportation Company, 39th Transportation Battalion, 26th Transportation Group, supported USMC operations in the Demilitarized Zone. The vehicle sits with other trucks in a fortified base camp.

Top left: "Devil Woman," based on a 1969 Kaiser-Jeep M54A2C, served successively in three companies of the 54th Transportation Battalion: the 669th, the 512th, and the 523rd. Four .50-caliber machine guns and an unidentified object under cover are visible atop the box. **Top right:** "The Midnight Angels" was later renamed "Stepping Wolf," a word play on the popular rock group Steppenwolf, and the vehicle remained with the 512th Transportation Company. The basic truck was a 1969 Kaiser-Jeep M54A2C, registration 05B58769.
Above left: "Black Widow," left, a 1969 Kaiser-Jeep M54A2C of the 523rd Transportation Company, 39th Transportation Battalion, sits next to "Satan's Li'l Angel" during Lam Son 719 in March 1971. "Black Widow" carried several spare tires in the rear for emergency

repairs. **Above right:** On the left, crewmen John Liblong and Mitch Reynolds of E-Battery 41st Artillery Quad-50 (MG) observe a gun truck from 8GP 27T pass them as they await a convoy heading up QL-19 thru the treacherous An Khe Pass in November 1970. The transportation battalion's gun truck sports two individual .50-caliber MGs allowing them to cover both sides of the road during an ambush. The quad-.50 crew also uses their M60 as protection against ambush. (Lynn Wood, NDQSA Archives)

"Black Widow," left, and "Satan's Li'l Angel" were photographed from a different angle at Fire Support Vandergrift during Lam Son 719, March 1971. The fronts of the hoods of both trucks were yellow, and details of their front ends were picked out in white paint.

Top left: Although the armored boxes of gun trucks offered gunners some protection from shrapnel and small-arms fire, it was not proof against all hits. "Satan's Li'l Angel," based on a 1969 Kaiser-Jeep M54A2C, was destroyed by an RPG in a 20 February 1971 ambush. **Top right:** The entry hole from the antitank grenade is under the "G" in "Angel." In the ambush, driver Richard Frazier was killed and gunner Chester Israel was wounded. The vehicle was written off and the armored box was utilized on "Proud American." **Above left:** "Vengeance is Mine," a gun truck based on a 1969 Kaiser-Jeep M54A2C, saw duty with the 512th Transportation Company, 27th Transportation Battalion. A whip antenna for a radio is present, and a vertical exhaust stack protrudes through the rear of the fender. **Above right:** "Vengeance is Mine" rests in a truck park in the mountains of the Republic of Vietnam. A strap in a partially lowered position secures the windshield armor, hinged at the bottom. A vision aperture is only on the driver's side of the windshield armor.

Top left: The armored box of "Vengeance is Mine" with plywood on the bed is seen from the cab. The inner armored plates were connected to the outer plates with a flat sill with recesses for storing .50-caliber ammunition boxes. **Top right:** A gunner standing next to a .50-caliber machine gun on a pedestal mount lights a cigarette in the armored box of a gun truck, with "The Midnight Angels" in the background. "The Midnight Angels" was assigned to the 512th Transportation Company. **Above left:** "Glory Stompers II," a 1969 Kaiser-Jeep M54A2 of the 512th Transportation Company, was the successor to the original "Glory Stompers," destroyed in an ambush in An Khe Pass. The name was a tribute to a 1967 motorcycle gang movie starring Dennis Hopper. **Above right:** Three crewmen pose in the armored box of "Iron Butterfly" of the 512th Transportation Company. The vehicle shared its name with a then-popular heavy-metal group, and the style of the lettering on the vehicle was a close match to the band's logotype.

Top left: The round opening in the rear of the right fender through which the vertical exhaust stack of "Iron Butterfly" originally protruded is visible. On M54A2 gun trucks, the original exhaust stack often was rerouted because it was in the line of fire of the machine guns.
Top right: "The Boss," based on a 1969 Kaiser-Jeep M54A2C, served with the 545th Transportation Company of the 27th Transportation Battalion. Over time, the lettering of the name on this vehicle went through several noticeable changes in font styles and coloration.
Above left: The crewmembers of "The Warlords," based on a 1969 Kaiser-Jeep M54A2C, take a break. The truck was assigned to the 597th Transportation Company, 27th Transportation Battalion. Two .50-caliber machine guns are visible atop the armored box. **Above right:** This gun truck probably from a transportation group out of Chu Lai, circa 1971, was used for convoy security along QL1 in ICORP / IFFV. The artwork shows two horse heads and the name "IRONHORSE" on a gold background. The Diesel or multifuel-powered 5-ton is armed with two M60 MGs and one .50-caliber MG. (Joe Kotarba, G-55th, NDQSA Archives)

Top left: "Brutus"—the 359th Transportation Company of the 8th Transportation Group gun truck in which Larry Dahl earned the Medal of Honor on February 23, 1971. A convoy of tankers had been ambushed. "The Creeper" and "Playboy" had all they could handle, and after losing all their tires, the NCOIC of "The Creeper" called for help. "The Untouchable," "The Misfits" and "Brutus" were dispatched from another convoy to their aid. After the brunt of the fighting, a VC threw an explosive device into the bed of "Brutus." Gunner Dahl shouted a warning to his crewmates, and dropped on the device. This heroic act ended his life, but saved theirs. **Top right:** Another view of "Black Widow," one of the gun trucks serving with the 523rd Transportation Company. This truck was probably based on an M54A2C. It had pedestal mounts and ammunition boxes for three .50-caliber machineguns, but the guns are dismounted in this photograph. **Above right:** Several Vietnam War gun trucks were named after the cartoon character Snoopy. This one, "Snoopy II," with the 444th Transportation Company, 27th Transportation Battalion, had a tall box with indentations. Bo Carlson painted the artwork on the box. **Above left:** In the cartoons, Snoopy's nemesis was the Red Baron, so naturally there was "The Red Baron" gun truck, attached to the 27th Transportation Battalion. Under cover in the armor body is a GE XM-134 Minigun. "To Snoopy with Love" is written on the iron cross.

This USMC gun truck, apparently based on an M54A2C, was equipped with a ring mount for a .50-caliber machine gun. The truck was unusual in that the armor box was positioned back from the front of the cargo bed. Two carriages are on the ring mount.

Top left: "Ace of Spades" of the 523rd Transportation Company, 54th Transportation Battalion, 8th Transportation Group, was armed with three .50-caliber machine guns and sported a prominent ace of spades playing card painted on each side of the armor box. **Top right:** On 16 February 1971, "Ace of Spades" crashed into a ravine along Route 9, killing its driver, Michael R. Hunter, and injuring gunner Jack Turner. Sections and parts of the armor box are scattered on the bank. The vehicle is thought to have been an M54A2C. **Above left:**

An M60-armed gun truck proceeds down a dirt highway among bicyclists and civilian cars in the Republic of Vietnam. The number 33 is painted on each rear-view mirror, and, in an unusual configuration, the outer side armor panels are tilted outward at the top. **Above right:** Perhaps the most famous Vietnam gun truck is "Eve of Destruction," an M54A2C of the 23rd Transportation Company that survives at the U.S. Army Transportation Museum, Fort Eustis, Virginia. The machine gun ammunition trays and cradles were painted red.

Top left: "Eve of Destruction" is viewed from the left side. This gun truck was armed with a twin .50-caliber machine gun mount toward the rear of the armored box and with two single-mounted .50-caliber machine guns toward the front end of the armored box. **Top right:** A soldier works on the front end of "Eve of Destruction." The front .50-caliber machine guns are fitted with canvas covers. At this time the truck had a vertical exhaust stack. This later was eliminated in favor of an exhaust tail pipe routed under the right fender. **Above right:**

The rear plate of the armored box of "Eve of Destruction" was inside the tailgate. The hinges of the tailgate were painted white, as were details of the towing pintle bracket. The "69" at the end of the registration number indicates the truck was a 1969 model. **Above left:** "Eve of Destruction" is being prepared for shipment to the United States in 1971. With considerable foresight, Capt. Don Voightritter, commander of the 523rd Transportation Company, persuaded army brass to ship the gun truck to the museum at Fort Eustis.

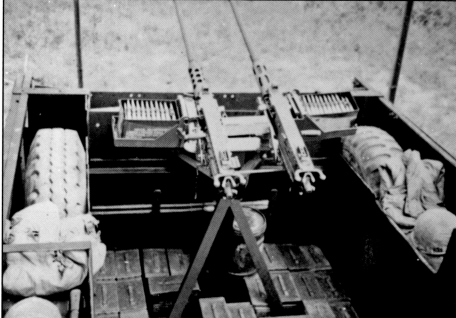

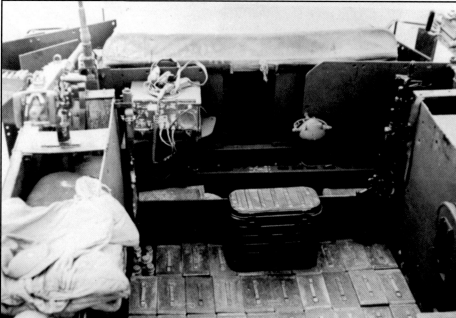

Left: The barrels of the twin-mount .50-caliber machine gun at the rear of the armored box of "Eve of Destruction" overhang the stowage space created by the inner (left) and outer armor plates at the rear of the vehicle. An M60 is stored on the inner plate, and ammunition boxes are stored in the space between the plates. The outer plate is positioned forward of the tailgate to the right. The spaced armor plates improved the ability of the armored box to defeat RPGs. **Top right:** The twin .50-caliber machine guns in the rear of "Eve of Destruction" were on a mount fabricated in the field from angle irons. The space between the outer and inner armor plates made a handy stowage space for spare tires, bedrolls, helmets, and other gear. **Above right:** At the front of the armored box of "Eve of Destruction" was a radio set. To the left is a .50-caliber machine gun. Fifty-caliber ammunition boxes covered the floor. An M79 grenade launcher is stowed next to the radio, and an M60 machine gun is opposite it.

Top left: In a view from the rear of the box of "Eve of Destruction," atop of the radio and connected to it with a coil-type cable is a flight helmet, sometimes used by gun truck crewmen because it had a built-in microphone and earphones. CVC helmets also saw use. **Top right:** Stored below "Eve of Destruction's" forward right M2 .50-caliber machine is an M60 machine gun and a helmet. On the tightly packed ammunition boxes on the floor is a Mermite insulated food container, ubiquitous on U.S. vehicles during the Vietnam War.

Above left: The right forward corner of the armor box of "Eve of Destruction" is viewed. An M60 machine gun is stored vertically at the center. Mounted on the inner side armor is a folding seat of the type sometimes used in U.S. tanks and armored vehicles. **Above right:** The cab of "Eve of Destruction" is viewed through the driver's open door. A large box, no longer present on the vehicle as displayed at the U.S. Army Transportation Museum, is on the passenger's side. Details of the armor on the passenger's door are visible.

"Eve of Destruction" rests alongside "Black Widow" at a base in the Republic of Vietnam. On "Eve," "To Charles With Love" is written on the bumper, along with markings for vehicle 214, 523rd Transportation Company, 39th Battalion, 8th Transportation Group.

In June 1971 "Eve of Destruction" is being hoisted aboard a ship in the Republic of Vietnam for transportation to the United States. The angled outer armor of the passenger's door and the revised exhaust pipe are quite apparent from this angle.

Top left: "The Untouchable," based on a 1969 Kaiser-Jeep M54A2C, was assigned to 359th Transportation Company, 27th Transportation Battalion, Qui Nhon Support Command (QNSC). The armored box had been salvaged from the gun truck "Ball of Confusion." **Top right:** "Only the Strong Survive" is painted on the rear of "The Untouchable." In addition to a .50-caliber machine gun, two General Electric XM134 7.62mm Miniguns are mounted in the armor box. What appear to be flak vests are draped over these guns.

Above left: A General Electric XM134 7.62mm Minigun mounted on a gun truck and its flex feed chute are viewed. The gun was mounted in a cradle fabricated from angle irons and rods and could be swiveled and elevated to cover most angles to the sides of the vehicle. **Above right:** A gunner lays down fire with a GE XM134 7.62mm Minigun. This weapon could fire at an astonishing rate of up to 6,000 rounds per minute—or 100 rounds per second—although gun truck gunners in Vietnam seldom fired these guns at their maximum rate.

This Minigun-armed gun truck had slightly curved upper rear corners on its side armor plates, each of which seem to have been fabricated from two plates with a horizontal welded seam. In addition to the Minigun, an M2 .50-caliber machine gun is mounted.

APC-bodied Trucks

As an expedient to produce gun trucks at a time when there was a shortage of steel plates in Vietnam, M113 armored personnel carriers (APCs) sometimes were shorn of their suspensions and placed on the cargo bodies of M54A2C 5-ton trucks. These conglomerations were known as APC gun trucks. One example was "King Kong" of the 523rd Transportation Company. APC gun trucks provided good protection but were not immune to RPGs; "King Kong" was penetrated by an RPG during Lam Son 719.

Top left: The dismounted body of "King Kong" sits in an equipment yard after it had been removed from one truck chassis. Later, it would be mounted on a different 5-ton chassis. An "Okinawa" style machine-gun shield is over the cupola, and the cargo hatch is open. **Top right:** The yellow band on the nose of King Kong's hood was a marking of the 8th Transportation Group, permitting the unit's vehicles to be readily identified from the air. The 8th Group arrived in Vietnam 19 October 1966, and departed 28 April 1971. **Above left:** "The

Big Kahuna" was an APC gun truck serving with the 512th Transportation Company. Hooks with turnbuckles attached to the lifting eyes on top of the M113 glacis helped to secure the APC body. "Wes" is stenciled under the driver's side vision port. **Above right:** As seen in a rear view of "The Big Kanuna," the M113 body was a good fit inside the cargo bed of the M54 5-ton truck. On each side of the top cargo hatch is an M60 machine gun. A ladder is hinged to the rear of the cargo bed to facilitate access to the APC body.

Top left: "King Cobra" was an APC gun truck assigned to the 597th Transportation Company of the 27th Transportation Battalion. Atop the APC body were three .50-caliber machine guns with ACAV shields. The exhaust stack is of an atypically tall model. **Top right:** "King Cobra" is observed from the right side, showing the angled glacis of the M113 body that was visible from this angle. The rims of the wheels were painted a light color, probably white. The inset door of the ramp at the rear of the M113 body is open. **Above left:** "King Cobra," center, rests in a truck park with another M54-based APC gun truck of the 597th Transportation Company, "Big Bad John," left. "Big Bad John" was assembled as an APC gun truck in October 1969 and was redubbed "Sir Charles" in January 1970. **Above right:** When this photograph of "King Cobra" was taken, its ACAV cupola shield had been repainted in a patriotic stars-and-stripes design. The two front armored plates of the ACAV shield were not present, and the .50-caliber machine gun is covered.

Top left: The former "Big Bad John" is shown rebranded as "Sir Charles" sometime after the beginning of January 1970. The truck was assigned to the 597th Transportation Company, 27th Transportation Battalion, now under the Qui Nhon Support Command (QNSC). **Top right:** Several crewmen relax on top of the M113 APC body of "Sir Charles" as observed from the left side. The light-colored artwork on the side of the body was a tombstone with an "R.I.P." heading and with a Vietnamese rice hat lying on top of the tombstone. **Above left:**

Two gunners pose for their photograph on top of the M113 body of "Sir Charles." To the left is the ACAV cupola shield, protecting the .50-caliber machine gunner. To the left, a gunner is poised with an M60. To the far left are the driver's vision blocks of the M113. **Above right:** A gunner is test firing a .50-caliber machine gun in the cupola of an APC gun truck. The gunner's right hand is visible to the far left. The gun protrudes through the frontal plates of an ACAV cupola shield. To the lower right is the helmet of a crewman observing fire.

"Set Me and Lt. Calley Free" was an APC gun truck of the 572nd Transportation Company. The artwork showed U.S. Army Lt. William Calley, who was tried for murder in the My Lai Massacre in 1971, behind prison bars. Drying laundry is in evidence.

An APC gun truck of the 2nd Transportation Company, 27th Transportation Brigade, accompanies a Jeep on a sweep of Highway 19 on 10 June 1970. The M113's mud flaps overlap the sides of the cargo body of the M54 truck: a common configuration on these vehicles. The armaments on the M113 body include three M2 .50-caliber machine guns: one in the ACAV cupola shield and one on each side of the top cargo hatch. (NARA)

Top left: Gun trucks, including several APC gun trucks, of the 27th Transportation Battalion are lined up. Toward the left are two ACAV-type cupola shields mounted atop M113 bodies that rest on the second and third M54s in line. The fronts of the hoods are yellow, typical of 8th Transportation Group. **Top right:** "Killing is our business and business is good!" reads the slogan on the rear of the M113 body mounted in this 563rd Transportation Company truck, operating out of Pleiku. Named "The Lifer," the side of the APC body was adorned with the Sergeant Snorkel cartoon character. **Above left:** In this instance, placing the APC body on a 1969 M54A1C, registration number 05B72969, has created the gun truck. The M54A1C featured a drop side cargo bed, immediately recognizable by the hinges along the lower edge of the bed, and was powered by a Mack ENDT-673 Diesel engine. **Above right:** "Colonel" was built on a M54A2, registration number 5E8261. M54A2 trucks did not have drop sides, hence the lack of the hinges along the lower of the bed as compared to the photo at left.

Showing a second paint scheme, "The Colonel" illustrates the evolving nature of the vehicles. The additional weight of the M113 body taxed the chassis of the 5-ton 6x6 severely. Worn, damaged or destroyed truck chassis were changed, paint schemes evolved, and crews and names changed in some instances. Men of the 363rd Transportation Company operated "The Colonel."

Engineer Gun Trucks

Engineer units often had to operate independent of combat units, making them prime targets for enemy attack or ambush. In response, the engineers soon began fabricating their own versions of gun trucks. The M328 5-ton bridge truck was a prime candidate for this. With the M4T6 bridge unloaded, the large flat bed was an ideal surface upon which a mobile fortification could be installed. These engineers have used galvanized culvert sections and a M49 ring mount in their creation. With an extra-long wheelbase, over width bed, and 14.00-20 tires, the gasoline-powered M328 was one of the largest tactical trucks fielded by the army—its imposing size evident as a team of trackers and their dogs from the 4th Infantry Detachment (War Dog Provisional) clamber into the shoulder-high truck bed. (NARA)

Top left: The considerable size of the M328 is apparent here, where men of the 553rd Engineer Float Bridge Company have created a gun truck by placing the hull of a M113 APC inside the bridge truck body, dubbing their creation "Road Runners." The truck bed side panels have been reversed, presenting a cleaner appearance, and the APC retains its full array of weapons and gun shields. **Top right:** Another M328/M113 hybrid was "VC Birth Control" "The Pill" created by the 20th Engineer Battalion out of Weight-Davis FB south of Pleiku in the Central Highlands. The bed side panels on this vehicle are in their standard orientation.

(Dale Stiebritz, B-4/60th Arty, NDQSA Archives) **Above right:** In Vietnam, some engineer units converted M51 dump trucks to gun trucks to provide security for construction, mine-clearing, and other operations. Here are "The Judge" and "Peace Maker," M51A2s of the 585th Engineer Company, 589th Engineer Battalion. **Above left:** A dump-truck-based gun truck, left, provides security for vehicles of the 18th Engineer Brigade, 937th Engineer Group at Pleiku, Republic of Vietnam. A .50-caliber machine gun is mounted in the dump body, the top of which is lined with corrugated metal.

Armored M151s

GI's in an armored M151 observe the aftermath of a response to an ambush along QL-19 in May 1968. The vehicle is fitted with a pedestal-mounted M60, and a large siren on the left front fender. (Rick Look, E-41st, NDQSA Archives)

Top left: Thin armor plates have been added to this Ford M151A1, registration number 2G6336, in the field. The new passenger's door consisted of a large lower plate with a smaller one bolted to the top. Separate plates protected the sides of the rear of the MUTT.
Top right: The same armored M151A1 had a low steel plate in the windshield position. It was reinforced with a brace on each side, and "MILITARY POLICE" was marked on the front on a light-colored background. **Above left:** A driver is in the armored M151A1, as seen from the left side, and the nickname "NICKEL" (or "NICKEL O"?) is painted on the

plate. The door was on the hinge, and it included a rough but apparently effective latch on the rear end that simply flipped down against the inside of the body metal. (MP History Office, three) **Above right:** While there is little doubt that locally sourced material was used to armor vehicles, most of the armor applied to US vehicles in Vietnam was manufactured to for this purpose. In the view the factory applied stenciling "Armor Module HH 7782633" is plainly visible, as are the stenciled illustrations indicating how the panels were to be assembled. (MP History Office)

A M151A1 "Mutt" of H-Battery 29th Artillery (SLT) is outfitted with the 23" Xenon searchlight and a gun mount for an M60 machine gun. When the searchlight is in operation at night in either infrared or white light mode, the position immediately becomes a target of enemy fire. The M60 can help protect the crew and light. (Joel Nelson, H-29th, NDQSA Archives)

A fairly sophisticated armor kit has been installed on this M151 of Company A, 720th MP Battalion, 18th MP Brigade, seen during a rest stop while escorting a truck convoy to Cu Chi in August 1967. The door openings have frames, and ballistic glass is on the front shield and the doors. The armor has been made to fit inside the radio-antenna support. (NARA)

Armored M37s

One of the smaller types of gun truck was that based on the Dodge M37B1 ³/₄-ton 4x4 truck. This example, "Daughter of Darkness," previously had gone by the name "Malfunction." It was assigned to the 523rd Transportation Company, 39th Battalion.

Top left: "Otto II," a gun truck based on a Dodge M37, served successively with the 64th Transportation Company and the 359th Transportation Company. Two pedestal-mounted M60 machine guns were in front of the armored box and one M60 was inside the box. **Above left:** This M37B1, probably a 1963 model, was outfitted as a rudimentary light gun truck with armor plates fastened inside the stakes of the cargo body. A folding armor plate to protect the windshield, hinged at the bottom, is installed. In the rear is an M60 machine gun. **Right:** A 7.62mm XM134 GE Minigun is mounted on an M37 truck about to depart on escort duty with a convoy from Fire Base Schroeder on July 20, 1969. Grips for directing the Minigun are visible. The weapon is on a large, U-shaped cradle on a pedestal mount. The vehicle is equipped with a radio (left) and an antenna. Note the electrical cable for powering the gun. (NARA)

The Last Survivor

As the United States ground forces wound down their involvement in the Vietnam War, scores of gun trucks were restored to cargo vehicles, scrapped, or transferred to the Army of the Republic of Vietnam. One gun truck that was preserved as such was sent back to the United States: "Eve of Destruction," an M54A2C that had served with the 523rd Transportation Company, variously assigned to the 8th Transportation Group's 39th, 54th, and 124th Battalions. In June 1971 "Eve of Destruction" was shipped from Vietnam to the U.S. Army Transportation Museum at Fort Eustis, Virginia, where the vehicle remains on display.

Top left: Surrounding the vehicle is a large display that recounts the service of the truck in Vietnam. **Top right:** "Eve of Destruction's" original exhaust stack was replaced by a new tail pipe below the fender. Below the passenger's door is the battery box with two 6TL 12-volt batteries wired in series to produce 24 volts. To the rear of the battery box is a toolbox. **Above left:** Mounted on top of the right fender of "Eve of Destruction" is an air-cleaner as-sembly. The air cleaner consists of an air intake at the rear, a canister containing the cleaner element, and, at the front, the air tube. The cleaner element was to be cleaned daily. **Above right:** The front end of "Eve of Destruction" is viewed. The truck wears bumper markings, left to right, for 8th Transportation Group, 124th Transportation Battalion, 523rd Trans-portation Company, 214th in the march order. Some details are painted white.

Top left: At the front of the chassis is a winch of 20,000-pound capacity. When used with a snatch block, the winch could be used to recover other vehicles or move heavy objects. Behind the winch is the lower front of the radiator. At the top right is a blackout driving light. **Top right:** As viewed from below and to the front, the front winch is of the worm-gear, jaw-clutch, horizontal-drum type with an internal automatic brake secured to the front of the worm-gear drive shaft, and is equipped with an internally mounted drum drag brake. **Above**

left: The front axle of the M54A2C was the Timken-Detroit Model FM 240-B2, a hypoid, double-reduction, full-floating type with a gear ratio of 6.443:1. It is secured to the frame with springs. Below the axle are cylinders to help take the weight off the suspension. **Above right:** The right side of the front suspension is viewed from the front. Attached to the leaf spring is a shock absorber. Two U-bolts secure each side of the axle to its spring. To the right is the spring hanger; a spring shackle secures the rear of the spring to the rear hanger.

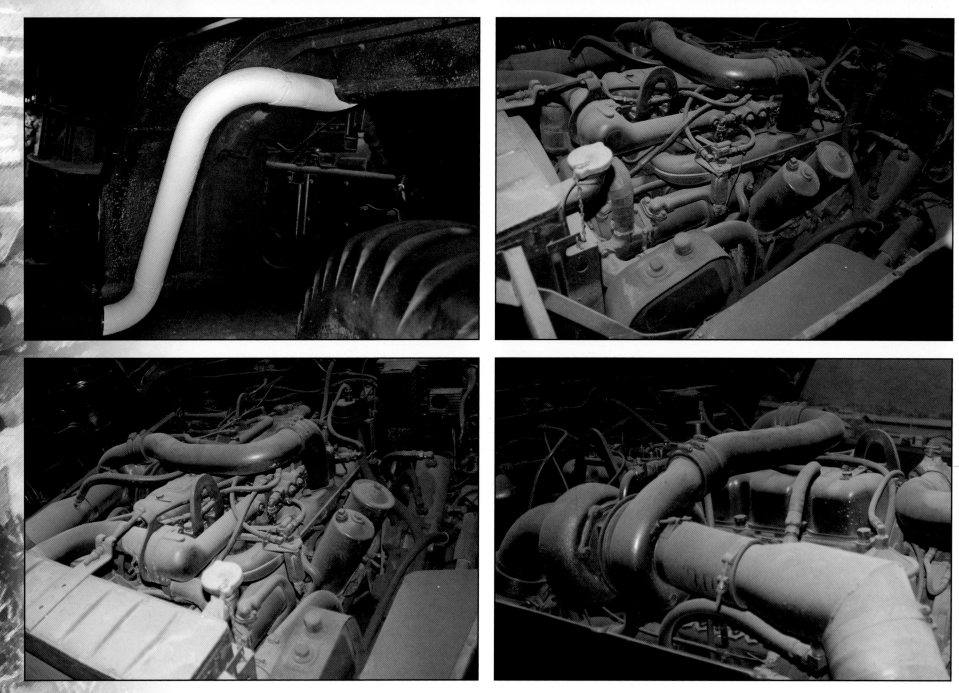

Top left: A view under "Eve of Destruction's" right fender reveals the modified exhaust tail pipe, which emitted engine gasses without use of a muffler. Above the bend of the tail pipe, the round port in the fender through which the original exhaust stack passed is plugged. **Top right:** "Eve of Destruction" is fitted with a Mack ENDT-673 diesel, which replaced the original Continental LDS-465-1A multifuel engine. The view is from the front left corner of the engine bay. The two tilted cylinders in the foreground are the oil filters. **Above left:** The

Mack ENDT-673 diesel engine is viewed from a slightly different angle. In the left foreground is the top of the radiator. At the upper right, mounted on the left of the firewall, is the generator regulator. To the upper left is the air-cleaner assembly. **Above right:** The engine bay of "Eve of Destruction" is observed from the front right corner. In the foreground is the carburetor-air duct, with an inline turbocharger of the oil-cooled, exhaust-driven type with a blower compressor visible toward the left foreground.

Left: The rear portion of the air cleaner on the right fender is viewed. The air intake on the rear of the air-cleaner canister apparently was a nonstandard type. The standard air intake on this unit was a downward-pointing elbow-shaped pipe with the air intake opening located on the bottom. **Right:** The frame for the passenger's side rear-view mirror is shown. It is of tubular construction. The mirror itself is out of view to the left. Toward the lower left is a mounting bracket for the frame on the cowl. Also in view is part of the armor for the passenger's door.

Left: The passenger's door of "Eve of Destruction" is open, allowing a view of the arrangement of the door armor. A large plate was applied to the exterior of the stock door, extending aft of the door. An inner plate is mounted atop the stock door. Both plates have viewing ports. **Right:** The leading edge of the armor plate on the outside of the passenger's door (here), as well as the driver's door, was screwed tight to the door, but the rear edge of the plate was spaced out from the door with two U-shaped brackets, allowing access to the exterior door operating handle.

Left: The space between the rear of the passenger's door, left, and the armor plate on the out-side of the door, right, is viewed. To the lower left is the exterior door handle, and at the center of the photo is the upper bracket attached to the outer armor and the armor plate above the stock door. **Top right:** With the passenger's door open, the inner plate of armor above the stock door and the outer armor plate (with the holes near its rear edge), extending well aft of the inner plate, are seen close-up. The vision ports were roughly cut through the armor with a torch. **Above right:** On the lower part of the interior of the passenger's door of "Eve of Destruction," the oblong object near the bottom is an access panel referred to as the "inside cover" or the "inspection plate" in technical manuals for the M54 5-ton 6x6 trucks.

Top left: As viewed through the passenger's door, "Eve of Destruction" has an armored windshield fitted with two bullet-resistant windows and windshield wipers. The original glass windows were removed. To the left is part of the armor on the driver's door. **Top right:** The interior of the armored windshield is observed close-up from the right. The bullet-resistant glass blocks are held in place by steel frames welded to the armor plate. The thickness of the glass can be seen through the gap at the bottom of the nearer frame. **Above left:** On the right side of the instrument panel is a glove compartment for storing documents and manuals. A simple latch secures it. Above the glove compartment are data plates with servicing data and a list of publications applying to this vehicle. **Above right:** At the center of the instrument panel is the instrument cluster, with, top row, left to right, the fuel gauge, speedometer/odometer, tachometer, and temperature gauge. On the lower row are oil-pressure and air-pressure gauges and the battery-generator indicator.

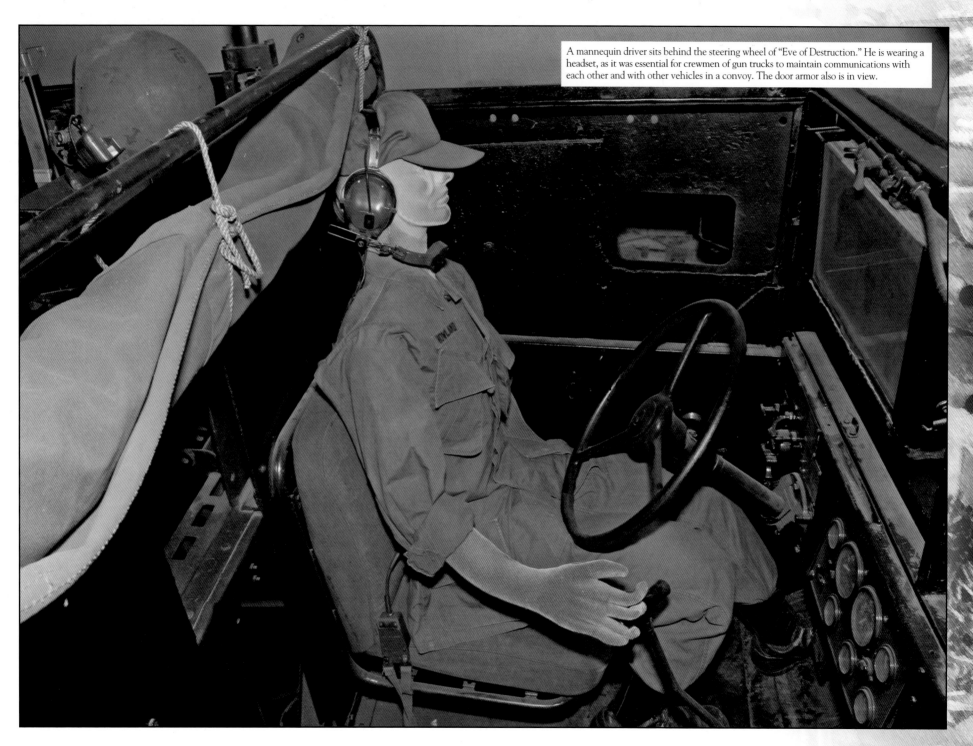

A mannequin driver sits behind the steering wheel of "Eve of Destruction." He is wearing a headset, as it was essential for crewmen of gun trucks to maintain communications with each other and with other vehicles in a convoy. The door armor also is in view.

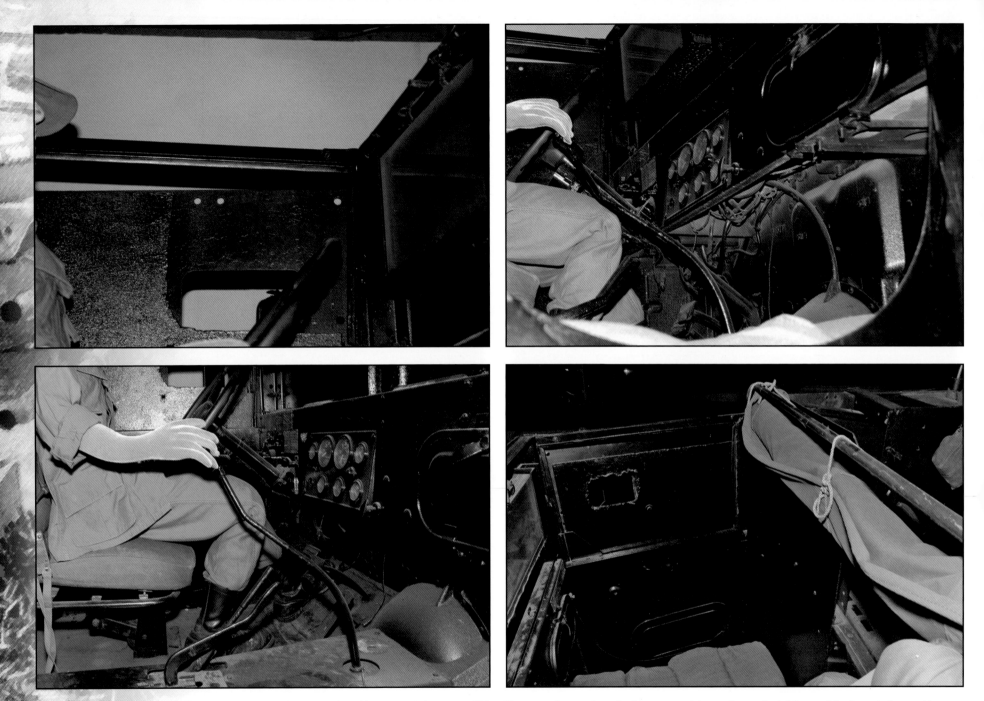

Top left: Whereas the interior plate of the passenger's door armor of "Eve of Destruction" is intact, the interior plate on the driver's side, here, has a chunk missing above and forward of the vision port. Note the roughness of the interior plate as compared to the exterior plate. **Top right:** The area of the cab below and to the front of the instrument panel is observed through the open passenger's door. The mannequin's hand is on the transmission gearshift lever. By the mannequin's leg are the transfer shift lever and the front winch control lever. **Above left:** The driver's seat consists of seat and back cushions with fabric covers on a tubular frame. Accelerator, brake, and clutch pedals are visible. **Above right:** In "Eve of Destruction," the passenger's seat was removed, and sandbags are arranged on the floor for protection against mines.

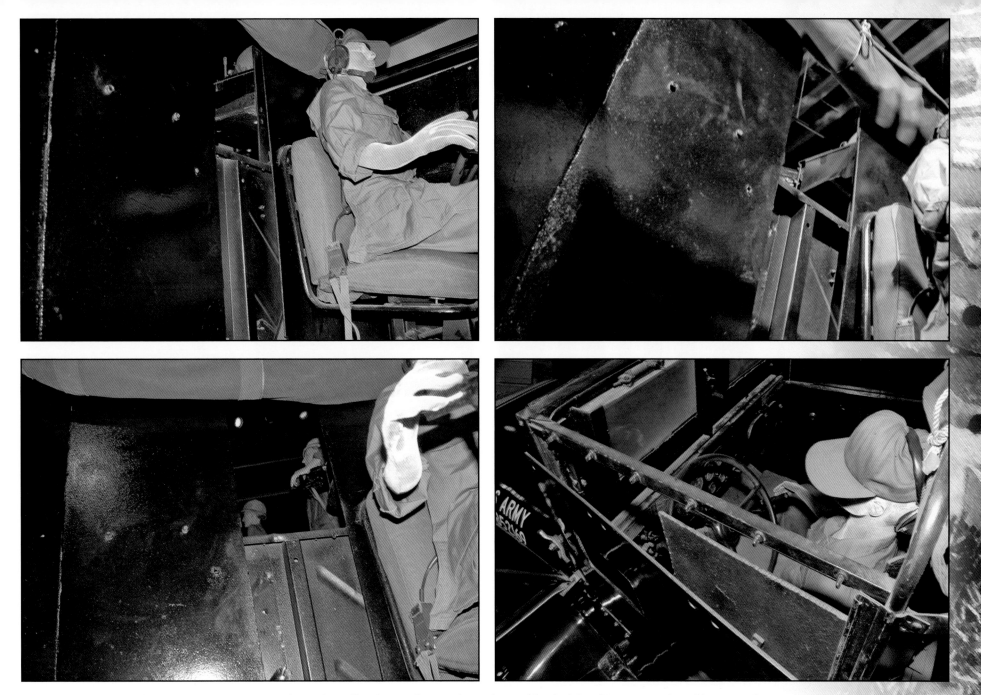

Top left: At the rear of the cab of "Eve of Destruction" are armor plates with a small gap between them. Kit armor generally came in $1/4$-inch and $1/2$-inch thicknesses; this appears to be $1/2$-inch. Above the mannequin's head is the rolled-up soft top for covering the cab. **Top right:** The armor plates at the rear of the cab are viewed from the passenger's side looking upward. The rear of the cab is visible in the gap between the two plates. This gap provided a means of communications between the driver and the gunners. **Above left:** The gap between the armor plates at the rear of the cab of "Eve of Destruction" is observed from next to the driver's seat. In the background, standing inside the armored box in the cargo body, are two mannequins representing the gunners. **Above right:** The right side of the cab of "Eve of Destruction" is viewed from above the driver mannequin. To the front is the left bullet-resistant windshield. To the front of the steering wheel on the left side of the steering column is the directional turn-signal control.

"Eve of Destruction" is painted black overall. Some gun trucks in Vietnam were painted black, reportedly in order to impart a sinister or menacing look to them. The lettering of the name is white with red shadowing. The rims of the wheels and the lug nuts are white.

"Eve of Destruction's" armor box originally was on an M54A1 but it was mounted on the current M54A2C after an RPG round damaged the vehicle in 1969. The M54A2C is a drop-side version of the cargo truck. The hinges of the drop panels are picked out in white.

Top left: The right dual tandem wheels of "Eve of Destruction" include white detailing on the rims, hubs, and lug nuts. The tires are size 11:00-20 with staggered, non-directional tread. The wheels are offset disk-type units with 10 lugholes and 5 oval ports. **Top right:** The right tandem suspension is visible between the tires. Below the center of the spring and the U-bolts is the spring seat-bearing cap. There are two torque rods below the spring on each side of the tandem suspension and two torque rods above the right spring. **Above left:** The left side of

the rear tandem axle on "Eve of Destruction" is seen from under the rear of the cargo body. The assembly on top of the axle is the differential carrier. On the bottom of the axle adjacent to the wheel is the rear of the left rear torque rod. **Above right:** The right side of the rear tandem axle is observed from the rear. The tandem axles are Timken-Detroit Model M-240-C3 with a gear ratio of 6.443 to 1.00. As is the case with the front axles, pipes are fitted under the rear tandem axle to help support the weight.

Top left: The right rear mudguard of "Eve of Destruction," viewed facing toward the rear, has a steel top and rubber flap. Two diagonal braces made of angle iron are attached to the guard. At the top are several hex screws that hold the side armor to the sill of the body. **Top right:** The rear of the right rear wheel well is observed. At the center is the rear of the right side member of the frame, of pressed-steel channel construction. Attached to the frame is a taillight assembly. At the top center of the photo is a crossmember of the frame. **Above left:**

Between the bumperettes at the rear of the frame of "Eve of Destruction" is the towing pintle, which is turned sideways. An artistic crewman in Vietnam painted the raised details of the towing pintle bracket white, and they remain that way today. **Above right:** On the left rear of the truck are a reflector, left, and a taillight assembly. They are mounted on a bracket on which the side holding the reflector is farther to the rear than the side with the taillight, which makes the rears of the light and the reflector virtually even.

The right rear corner of "Eve of Destruction" is viewed. The white hinges of the side drop panels and the tailgate is readily discernible. Clear plastic jackets are fitted over the tailgate chains. The rear edge of the rearmost side armor plate is fairly jagged.

Top left: The left side of the armor box of "Eve of Destruction" is observed. With the exception of the rear armor plate, the plates are held in place with a white hex screw at each corner. The "TP 70" stencil at the bottom of the second plate from left refers to tire pressure.

Top right: Parts of the two armor plates at the rear of the left side of the armor box are displayed. On the edges of the plates are unused screw holes arranged in pairs. These were predrilled in the kit armor since in some cases it would be necessary to place screws through the holes. **Above left:** As seen on the front mudguard of the tandem suspension, the steel upper part of the guard has V-shaped stiffeners. The rubber flap is fastened to the bottom of the steel guard with screws, and the bottoms of the braces are fastened on top of the rubber flap.

Above right: The fuel tank is forward of the mudguard shown in the preceding photo. The tank is secured to its brackets with straps, which are painted white. On top of the tank are a filler cap and a vent. The forward side of the tank is stamped with an X-shaped stiffener.

Top left: The right hand side air cleaner assembly. **Top right:** The menacing front grille work of the M54 series is represented here. "To Charles with Love" is inscribed across the bumper—in reference to "Victor Charlie," or the Viet Cong. **Above left:** Mounted on the fender on a combination bracket and brush guard is the left blackout marker/turn signal/parking light assembly. The vertical pole on the left side of the bumper helped the driver visually estimate the bumper's clearance with objects. **Above right:** The front left wheel and tire is viewed close-up. The lug nuts are painted white. Torque on these nuts was to be set at 450 to 500 lb.-ft. Tire pressure was 70 psi for highway use and 35 for cross-country. Inboard of the wheel, above the spring, is the lower drag link.

Top left: Details under the left fender of "Eve of Destruction" are displayed. Below the fender to the extreme left is the Pitman-arm shaft of the steering gear. Through the opening between the rear of the fender and the frame of the truck are the two air brake reservoirs. **Top right:** "Eve of Destruction" is observed from next to the front left fender. Stenciled in Vietnamese on the driver's door is a warning that it is dangerous to bicycle close to the truck.

Enemy terrorists posing as innocent bicyclists sometimes attacked convoys. **Above left:** A look backwards at the armor on the driver's door. **Above right:** The underside of the chassis of "Eve of Destruction" is viewed from under the driver's-side running board to the rear. Elements of the drive train are in view, from the transmission to the top left to the transfer case, drive shafts, and tandem axles.

Three mannequins man the machine guns in the armored box of "Eve of Destruction," as viewed from the rear of the vehicle. In the foreground is the spaced armor at the rear of the box, as documented in an earlier vintage photo of this truck. The ammunition box holders of the twin 50-caliber machine guns are painted red, as they were during the Vietnam War, but at that time the cradles and ammo box holders of all of the .50-caliber machine guns also were painted red.

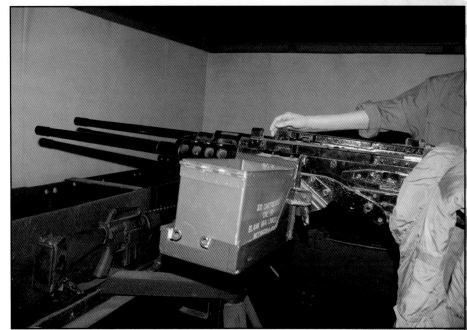

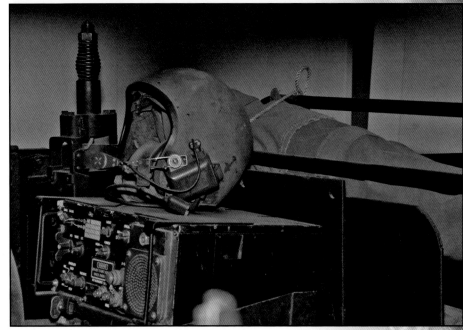

Top left: The right .50-caliber machine gun on the twin mount has the type of cooling sleeve with elongated openings, while the left gun has round openings in the sleeve. Vintage photos from Vietnam show that this was the correct configuration for the guns on this truck.
Top right: The cooling sleeve with round openings on the left machine gun of the twin mount in the rear of "Eve of Destruction" is shown just above the ammunition box. The ammunition box had a capacity of 100 rounds. An M16 rifle is stowed on the armor plate.

Above left: The right forward .50-caliber machine gun is on a pedestal mount, the pedestal passing through a horizontal steel plate between the outer and inner armor plates. A stowed M60 machine gun is in the foreground, and a spotlight is on the armor panel at the center.
Above right: A combat vehicle crewman's (CVC) helmet lies on top of the radio in the front of the armor box of "Eve of Destruction." Many gun truck crewmen preferred this type of helmet because of its built-in radio/intercom headphones.

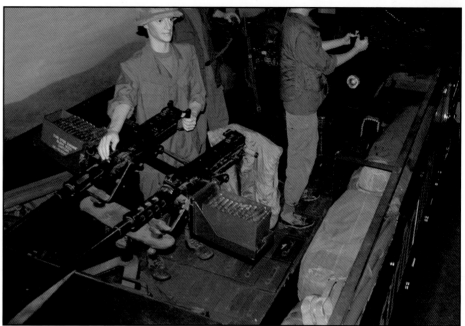

Left: The twin .50-caliber machine gun mount is observed from the right rear corner of the armor box. The pedestal that supports the gun mount is braced with a support frame fabricated in the field and consisting of angle irons. **Top right:** A Mermite insulated food container is also present, behind the gunner on the right. **Above right:** A flak vest is draped over the left machine gun in the twin mount. A clear view is provided into the space at the rear of the armored box between the outer and inner armor plates. A spare tire is stored between the inner and outer plates on the left side of the box. An M16 is stowed on the inside of the plate directly below the twin .50-caliber machine gun mount .

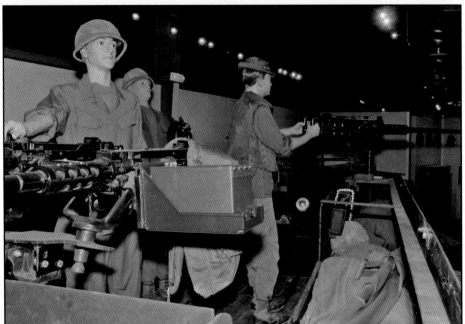

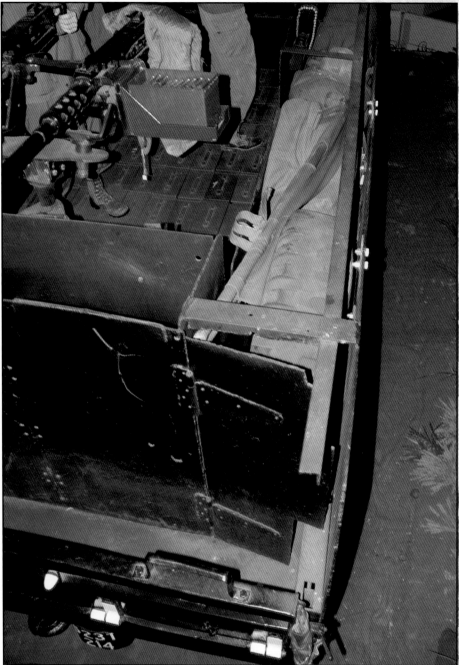

Top left: The manner in which the rear armor plates of the box of "Eve of Destruction" are spaced forward of the stock tailgate of the truck is illustrated from this angle. The outer panels of the rear of the box are hinged, apparently for ease of loading. **Above left:** The twin .50-caliber machine guns are on individual pintle mounts with ammunition trays and boxes on the outboard sides. These mounts in turn were installed on a swiveling plate on top of a pedestal so the guns could be traversed in unison from side to side. **Right:** The storage area created by the space between the outer and inner side armor plates on the right side of "Eve of Destruction" contains a spare tire. This space also formed a convenient place to stow bedrolls, tarpaulins, helmets, and other vehicular and personal equipment. The door at the right rear of the armor box is slightly ajar.

The twin .50-caliber machine gun mount at the rear of "Eve of Destruction" is viewed close-up. Ammunition boxes are still tightly packed on the floor of the armored box, as they were during the Vietnam War.